COAL IN QUEENSLAND

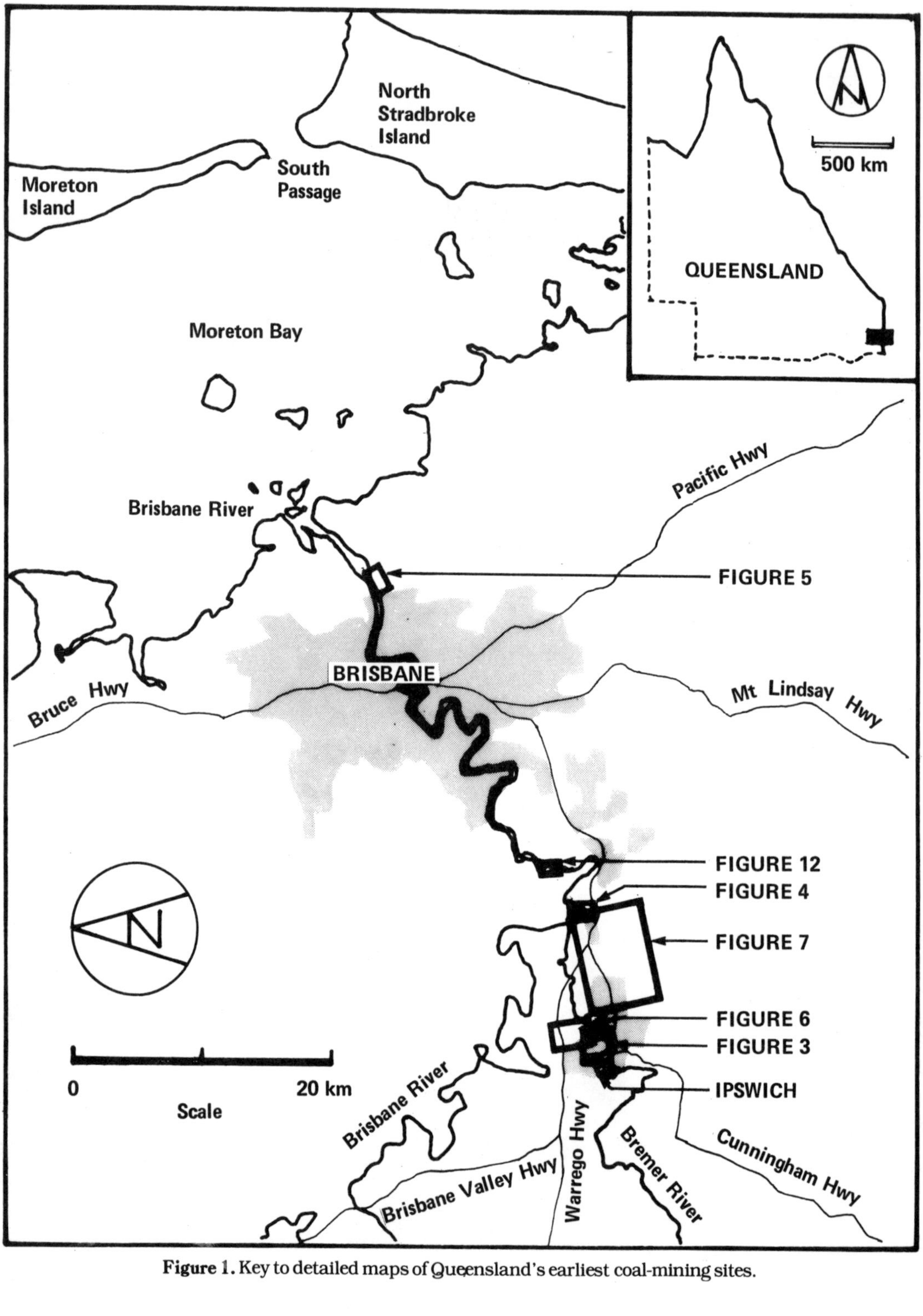

Figure 1. Key to detailed maps of Queensland's earliest coal-mining sites.

COAL IN QUEENSLAND

The First Fifty Years

A history of early coal mining in Queensland

R. L. Whitmore

University of Queensland Press
St Lucia • London • New York

Typeset by Savage & Co. Pty Ltd, Brisbane
Printed and bound by Silex Enterprise & Co., Hong Kong

Distributed in the United Kingdom, Europe, the Middle East, Africa, and the Caribbean by Prentice-Hall International, International Book Distributors Ltd, 66 Wood Lane End, Hemel Hempstead, Herts., England.

Published with the assistance of the Queensland Coal Owners' Association.

National Library of Australia
Cataloguing-in-Publication data

Whitmore, R. L. (Raymond Leslie), 1920 –
Coal in Queensland the first fifty years.

Includes index.
ISBN 0 7022 1619 4.

1. Coal trade – Queensland – History. 2. Coal mines and mining – Queensland – History. I. Title.

338.2′724′09943

Contents

Plates

Documents

Figures, Maps and Diagrams

Table

Note on Measures

Metric measures have been generally used throughout this book. However Imperial measures have been used where they occur in quotations or are appropriate because of common usage. Where it is felt necessary to give an indication of money values in contemporary Australian currency the conversion has been made at the rate of £1 = $2.

Conversion Factors

Money	£1 = 20s = 240d = $2
Length	1 yard = 3 feet = 36 inches = 0.9 metre; 1 chain = 22 yards; 1 mile = 1.6 kilometres
Area	1 acre = 0.4 hectare
Weight	1 lb = 0.45 kilogram; 1 ton = 20 cwt = 1 tonne
Volume	1 gallon = 4.5 litres
Power	1 horsepower = 0.75 kilowatt

The approximations are chosen to optimise simplicity of conversion with the appropriate precision of the measurement.

Introduction

A taste for history is a sign of maturity in a society. The past invariably influences the present to some degree and the significance of today's actions often comes into full focus only when seen against yesterday's events. Queensland's non-Aboriginal history now extends over a period of 150 years and it is, therefore, hardly surprising that Queenslanders are starting to look back with rising curiosity at the complex interplay between people and past events which has shaped their present society. No activity in the state has had a more varied or checkered history than coal mining, and a careful study of its origins is overdue.

After a meteoric climb in size and productivity over the last decade, coal mining has become Queensland's largest and wealthiest industry. The world shortages in coking coal and energy suggest that it will continue to hold this position for many years to come. But coal mining was also one of Queensland's first industries, albeit a very humble one, and the story of its early days provides a fascinating glimpse of a community intent on establishing its own way of life in a new environment.

In the following chapters the mining of coal is traced from the first sightings on the banks of the Brisbane River within a year of the establishment of a settlement on Moreton Bay, through the granting of independent status to Queensland in 1859 to the opening of the railway from Ipswich to Brisbane in 1874. The events which occurred in that period can only be appreciated in the context of the prevailing economic and social climates and these matters are therefore also considered. It must be conceded immediately that the picture which emerges is incomplete and contains important gaps and omissions. This is to be regretted, but the paucity of the contemporary documentation, the dispersal and loss of papers over the last century, and the ravages of the climate on the sites of the early operations, combined in many cases with re-utilization of the land, make it inevitable. It is a significant and disturbing fact that less is known of the early days of coal mining in Queensland than of mediaeval gold mining in Wales for instance, or lead mining in Derbyshire. The present work may go a small way towards rectifying this state of affairs.

Over the years, tales of the early days of colonization have been handed down from generation to generation by word of mouth or in written articles, during which subtle but often important changes in the significance of the events described have occurred. An example is the first mining of coal in Queensland. Who can claim the credit for this distinction? Should it go to the explorer who first recognized a coal seam and consciously collected a sackful for test? Or to the blacksmith or lime burner who dug sufficient out of a local outcrop to satisfy his needs? Or to the entrepreneur who first offered coal for sale commercially? All might reasonably press their claims. In the following chapters an attempt has been made to establish priorities in these various fields of activity and to relate them to the rival claims which have been made from time to time.

A great deal of use has been made of contemporary documents, particularly government papers and the newspapers. The reader may occasionally feel that insufficient weight has been given to later accounts of mines and their operations but these omissions occur only where the writers have relied heavily and obviously on earlier material which is still available. It is a pity that in the majority of cases later writers have not acknowledged the source of their material, and in the present work a special effort has been made to identify the source of all factual statements. This may not be of interest to the general reader but it should assist those who wish to pursue any particular aspects further. Of course it is recognized that contemporary accounts are not necessarily factually correct—notably if they appeared in a local newspaper—but cross-checks have been applied wherever possible.

Grateful thanks are due to many people for their assistance in preparing this text. Special acknowledgments must be made to the officers of the Department of Mapping and Surveying and the Titles Office of the Queensland government for giving access to old records and maps, and for providing expert interpretations of them; to the State Archivist, Mr Paul Wilson, for his interest and assistance in locating documents and his help in analyzing them; to Mr Spencer Broughton and students of the

Department of Surveying in the University of Queensland for their surveying work and to the officers of the Mines Department for useful discussions and visits. Mrs Betty Crouchley and Mrs Helen Gregory provided valuable information on Moggill Mine.

Chapter 11 relies heavily on the painstaking research of Mr David Bailey and I am deeply indebted to him for permission to use it, while assistance in ground survey work by the Historical and Industrial Archaeological Research Group—and especially Mr Ian MacPhail—is gratefully acknowledged. Special thanks go to Mr Mark Whitmore for help with the photographic work and to my wife for her devoted and untiring assistance in typing and editing the text.

Finally, full acknowledgment is made to the Australian Heritage Commission for financing the research and to the Queensland Coal Owners' Association for their sympathetic encouragement to complete the work and their generous financial support at the publication stage.

The Early Sightings

1

As a result of the favourable report which Governor Brisbane received from John Oxley, the surveyor-general of New South Wales, on his return from the 1823 expedition to Moreton Bay it was decided to establish a convict settlement there. Oxley was in command of the government brig *Amity* when it left Sydney on 1 September 1824 with an advance party of fourteen soldiers and about thirty convicts for the new settlement. Also in the party were Allan Cunningham, the king's botanist, and Robert Hoddle, a surveyor.[1]

Leaving the commandant, Lieutenant Henry Miller, with the problems of establishing a habitable settlement on the present site of Redcliffe, Oxley and Cunningham set off in two boats on 17 September to explore the Brisbane River. About a mile upstream from the present Colleges Crossing they established a base camp and made excursions on foot to conspicuous local look-out points. In the diaries of these expeditions neither explorer mentions seeing any coal, although they do record the presence of other minerals such as the gravels in the bed of the river, and granite and sandstone on the bank.[2] This is not really surprising because Oxley was mainly interested in mapping the area through which the river passed, while Cunningham's interests were predominantly botanical. Moreover at no point would they have passed an obvious coal outcrop on the river. On the other hand, Meston stated that "Coal was first noticed by Oxley in the Brisbane River in 1824 . . . "[3] but he did not enter any details of the supposed discovery.

Almost a year later, Major Edmund Lockyer visited the Brisbane River under instructions from Governor Brisbane (plate 1). He had arrived in Sydney with the 57th Regiment some three months earlier[4] and was given orders to investigate a report from John Gray, the pilot of Port Jackson, that he thought he had seen a tribe of white men at the mouth of a creek near the present Ferndale Bridge on the upper reaches of the Brisbane River.[5] Lockyer was also to report on the animals, birds and min-

Plate 1. Major Edmund Lockyer. (Courtesy Royal Historical Society of Queensland.)

erals of the region,[6] although his special qualifications for this mission are unknown, his chief experience for the previous twenty years having been that of a regular officer in India and Ceylon.[7] On 10 September 1825 he set off up the river and three days later at a point above that previously reached by Oxley he reported in his diary having passed " . . . a very rapid fall—coal bed." On his return journey he stopped at the coal bed, collected a sample in a sack, and commented that there was " . . . fine coal just below". A map illustrating Lockyer's discoveries was published a couple of years later in London by Cross,[8] and a second map which is very similar in many respects is located in the Queensland Government survey office in Brisbane.[9]

However, the exact location of this first discovery is not immediately obvious. In the first place it is most unlikely that the map published by Cross in 1827 was either drawn or checked by Lockyer who was leading an expedition from Sydney to King George Sound on the south west coast of Australia in 1826 and 1827. Indeed the cartographer was careless enough to give Lockyer's name the wrong initial (*F* was used instead of *E*)[10] and on both copies the sinuous nature of the Brisbane River is very imperfectly drawn (figure 2). Cunningham in his 1828 report to

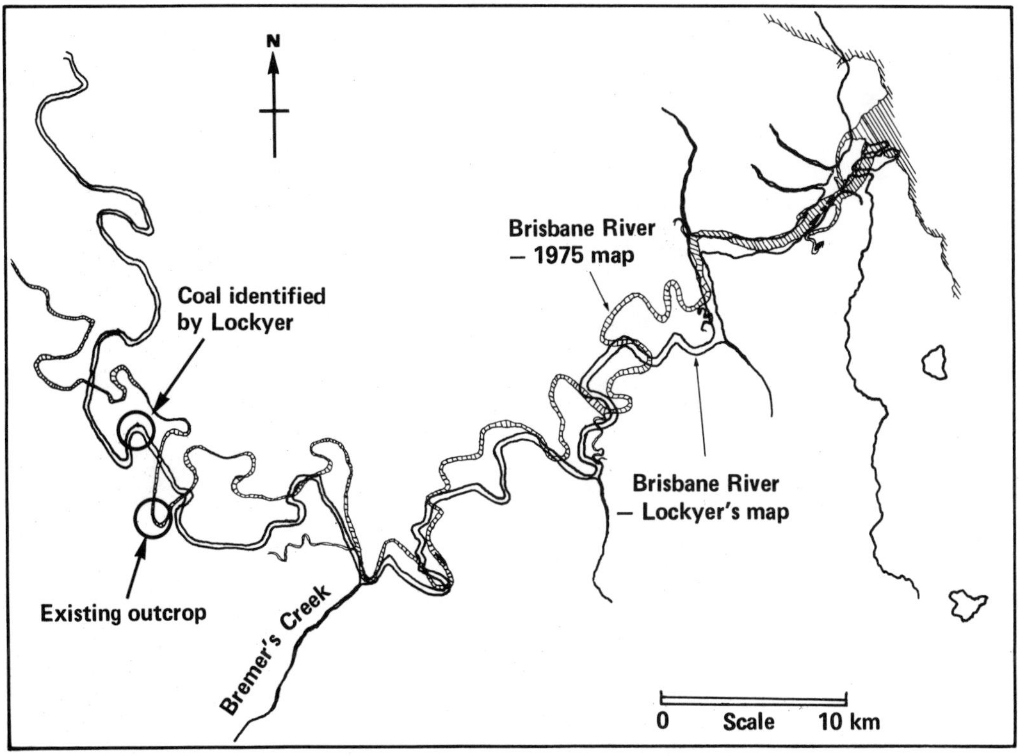

Figure 2. Lockyer's survey of the Brisbane River (chap. 1, n. 8 & 9) with a 1975 survey superimposed on it.

Governor Darling castigated Lockyer for his poor surveying and decided that

> . . . as he has not furnished us with a single bearing of any one known fixed point, I have left the river just where my late friend Mr. Oxley did, rather than add to it the trace of its chan-

> nel by Major Lockyer, which I observed marked on a badly engraved chart of the Colony recently published by Cross.[11]

Based on a comparison of the number and direction of bends traversed above the Bremer River junction, the coal bed ought to have been located some three miles above the present position of the Mount Crosby weir. The same conclusion was reached by N. Lockyer,[12] another investigator of the same name and grandson of the gallant Major. At about this point Cross's map[13] shows coal located on the outside of a right-hand bend but the survey office map[14] marks it on the inside of the same bend. Both positions are equally unlikely; the rocks in the area are lower palaeozoic and consequently not coal bearing so that there is no

Plate 2. Probable site of the first discovery of coal in Queensland. (Photographed in June 1980, 300 metres north of Coal Creek.)

question of a seam being visible on either side of the bend at the marked position. It is true that the inner bends of the Brisbane River possess large alluvial gravel beds and the possibility cannot be overlooked that coal pieces could have been broken off exposed seams now known to exist a little higher up the river and have been carried down by the water current for deposition in the gravel beds. In Lockyer's account of his expedition published in the *Sydney Gazette*[15] he specifically refers to a "vein of coal" which he observed, and not to a bed of gravel containing pieces of coal. A study of the distances and compass bearings which Lockyer recorded between the camping places on 12 and 13 September 1825 strongly suggests that the true place of his discovery was situated above the present Kholo Crossing, on the right-hand bank near the site of the weir and pumping station of the original Ipswich waterworks. Here the lowermost component of the Tivoli formation—the Waterworks Seam—outcrops quite dramatically for a distance of almost a kilometre immediately upstream from Coal Creek (plate 2). At this point the seam is 10 metres thick but interspersed with dirt bands.[16] A grab sample from the outcrop taken in 1977 by the author assayed 21.3 per cent ash which is about the average for coals currently mined in the Ipswich coalfield.

No further reference to coal has been found in the history of Queensland until after the arrival on 1 July 1828 of Charles Fraser at Brisbane Town, where the original encampment at Redcliffe had by then been moved. He was the colonial botanist and was accompanied by Allan Cunningham who had returned to Queensland in the hope of adding to the botanical collection at Kew in London.[17] On 11 July the two set out with the commandant of the settlement, Captain Logan (who had been posted to Brisbane Town in March 1826) on an expedition to Flinders Peak.[18] They were rowed up the Brisbane and Bremer rivers to the "Limestone Station" (now Ipswich) and Fraser noted that "Numerous beds of coal, lying in veins of considerable thickness, are adjacent to the lime; they jut out from the banks of the streams and fall into the Bremer within a few yards of its tide mark".[19] On returning from Flinders Peak, Fraser reported that the party had examined the coal seams which were abundant.

Although Cunningham no doubt joined Fraser and Logan in studying the coal seams during their joint expedition he makes no mention of them in his diary until he was setting out from Limestone Station a few weeks later on 18 August 1828 to explore a route to the Darling Downs through Spicer's Gap. He then describes how

> A stratum or seam of coal has been observed on the Bremer, both immediately above and below the Station, and as that mineral was noticed three or four miles to the north, in the steep banks of a dry creek dipping to the Brisbane, and again in another mile in the bed of that river, it is highly probable that the seam extends nearly horizontally throughout.[20]

On *Sketch B* which accompanied the report to Governor Darling, and which gives the date of the mapping excursion as 1 September 1828, coal is marked as located on the outside of the right-hand bend below Limestone Station, and also immediately above the Station (document 1). The former place corresponds to the mouth of the present Tivoli Creek. North of the Station and between the Bremer and Brisbane rivers the same sketch shows "Coal in fragments obtained in dry creeks".[21]

Over a period of years the legend has arisen[22] that the first person to discover coal in Queensland and, more

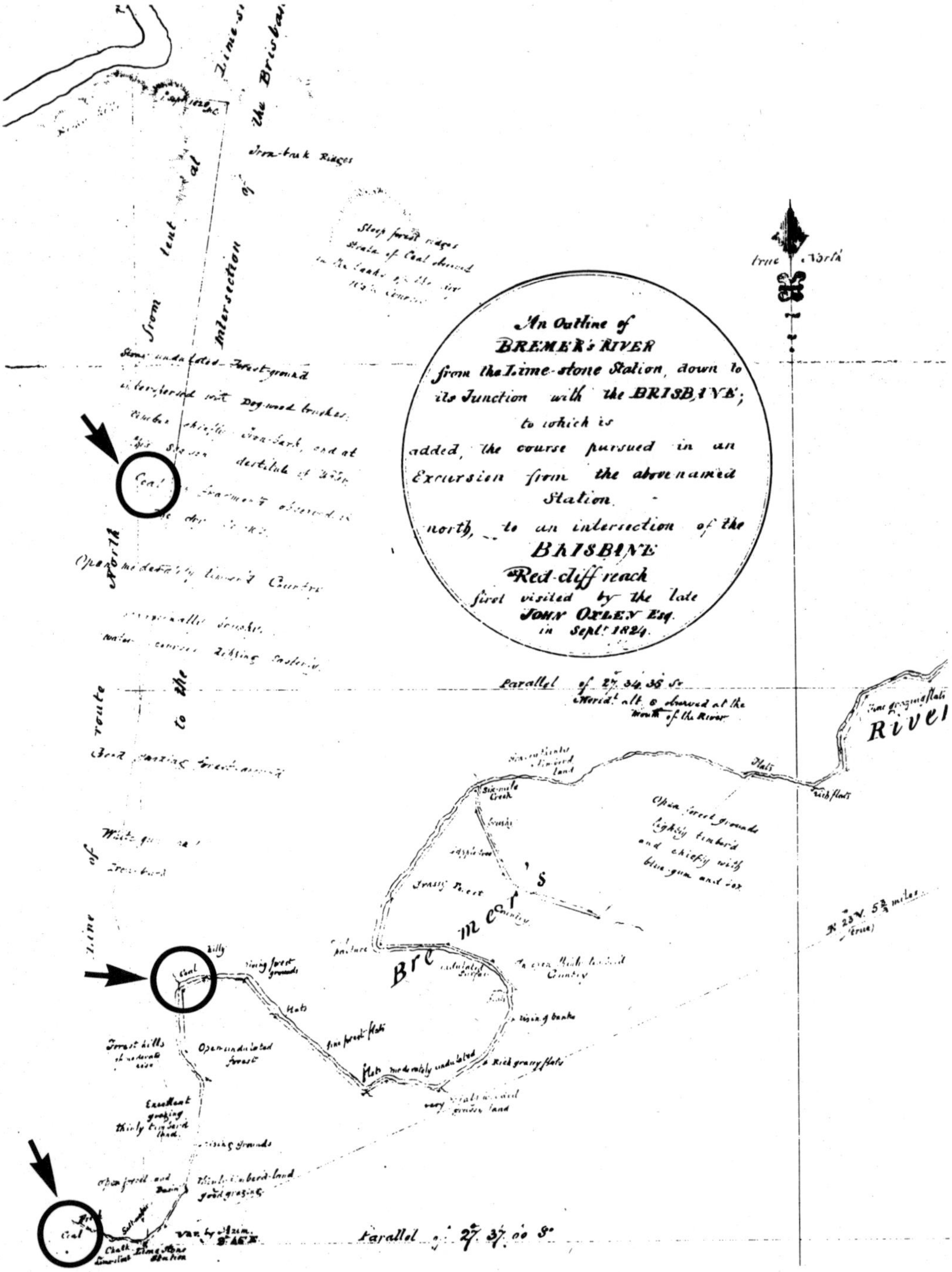

Document 1. Parts of Cunningham's 1828 map of the Bremer River (chap. 1, n. 21), showing locations of coal. (Plan supplied by the Surveyor-General, Queensland, and reproduced by arrangement with the Queensland Government. CROWN COPYRIGHT RESERVED.)

specifically, to find it on the Bremer River[23] was Captain Logan. A precise date for the discovery—8 June 1827—has also been mentioned[24] and the event is immortalized in the stone frieze which graces the Forgan Smith building of the University of Queensland (plate 3). Although these claims must inevitably fail because Logan did not arrive to take up duty as commandant of the penal settlement at Moreton Bay until six months after Lockyer had departed, the basis for the legend is worthy of examination.

Plate 3. Original drawing by Hennessy Hennessy Co., depicting Logan discovering coal in Queensland, for the stone frieze decorating the Forgan Smith building, University of Queensland.

Patrick Logan was a very enthusiastic explorer, and shortly after his arrival in Brisbane Town he led an expedition which discovered the Logan and Albert rivers.[25] But it was fifteen months after his arrival that he turned his attention to the Bremer River. On 7 June 1827 he set off by boat up the Brisbane and Bremer rivers camping on the first night at "the limestone hills".[26] The next day he sent the boat back to Brisbane Town and continued on foot. Whether he was the first white man to penetrate so far up the Bremer River is not stated in his diary but the fact that he gives no details of "the limestone hills", nor of the boat journey to them, suggests that they had already been discovered in the twenty-one months which had elapsed since Lockyer had mapped the junction of the Bremer with the Brisbane River. Unfortunately the journal of the expedition, which was forwarded to the colonial secretary on Logan's return to Brisbane Town, appears to have survived only as a newspaper article and could have been abbreviated. This published account makes no reference to Logan discovering coal on "the limestone hills" on 8 June 1827 or any other place at any other time. Nor do his biographers[27] make the claim on his behalf.

When Cunningham was resting at "Limestone Hills" in the August of the following year he recorded that "In the course of the last year, Captain Logan . . . discovered . . . Limestone Hills",[28] and this can be interpreted as meaning that Logan established the lime-burning community there in the second half of 1827. However Cunningham makes no mention of who first found the coal in the area, and the claim that it was Logan must therefore remain speculative until further evidence is forthcoming.

Despite the evidence that coal was known to exist close to Brisbane during the whole of the time that it was a convict settlement the later belief that it was being regularly mined or burned in those early days[29] is hardly confirmed by contemporary evidence and this is not really surprising. The internal economy of the settlement relied on wood for its fuel supplies, and it was cleaner and pleasanter to burn than coal. In his journal, Fraser had commented on the great profusion of hoop pines, yellow wood, tulip wood, figs and silky oaks between the settlement and Oxley Creek.[30] These would have been more than enough to provide heating and cooking for the settlement, to fuel the brick kilns and to produce the charcoal for use in the forges. The manufacture of burnt lime in the settlement from sea shells and at Limestone Station from chalk or limestone would have been equally satisfactorily fuelled with wood as with coal.[31] Indeed, in many parts of Queensland such a procedure was employed until well into the forties of the present century.

In 1828 the Australian Agricultural Company was granted the exclusive privilege for supplying coal to the public in the colony and it centred its operations on Newcastle, thus removing any potential competition from other sources. So the coal stayed in the ground, easy to mine and close to the surface, awaiting the time when it could be put to use and realize the potential which Lockyer, Logan, Fraser and Cunningham had obviously recognized for it in their early explorations.

Notes and References for Chapter 1

1. J.G. Steele, *The Explorers of the Moreton Bay District, 1770–1830* (St. Lucia: University of Queensland Press, 1972), p.127.
2. Ibid., pp.161–62.
3. A. Meston, *Geographic History of Queensland* (Brisbane: Government Printer, 1895), p.159.
4. *Australian Dictionary of Biography, 1788–1850*, Vol. 2 (Carlton: Melbourne University Press, 1967), pp.123–24.
5. E. Lockyer, "Journal of an Excursion up the River Brisbane in the Year 1825", *Australian Quarterly Journal* 1, ed. C.P.N. Wilton. See also H.S.Russell, *The Genesis of Queensland*, Appendix C (Sydney: Turner and Henderson, 1888), p.589.

6. *Australian Dictionary of Biography, 1788–1850*, Vol.2, pp.123–24.
7. N. Lockyer, "Exploration by Major Edmund Lockyer of the Brisbane River in 1825", *Journal of the Historical Society of Queensland* 2 (1935): 54–73.
8. Chart of Part of New South Wales (London: J. Cross, 1827).
9. Queensland Government Department of Mapping and Surveying, Brisbane, Map MT50.
10. Chart of Part of New South Wales.
11. Lockyer, "Exploration of the Brisbane River", pp.54–73.
12. Ibid.
13. Chart of Part of New South Wales.
14. Queensland Government Department of Mapping and Surveying, Map MT50.
15. "The Brisbane River", *Sydney Gazette*, 20 November 1825.
16. A.K. Denmead, "The West Moreton (Ipswich) Coalfield", *Queensland Geological Survey*, Publication 279 (Brisbane: Government Printer, 1955), p.35.
17. Steele, *Explorers of Moreton Bay*, p.225.
18. C. Fraser, "Journal of a Two Months Residence on the Banks of the Rivers Brisbane and Logan, on the East Coast of New Holland", *Botanical Miscellany*, Vol.1, ed. W.J. Hooker (London: John Murray, 1830).
19. Ibid.
20. H.S. Russell, *The Genesis of Queensland* (Sydney: Turner and Henderson, 1888), p.146.
21. Queensland Government Department of Mapping and Surveying, Map MT56.
22. Two authors who claim that Logan was the first to discover coal in Queensland are W. Roach, "Some Aspects of Coal Mining in Queensland", *Queensland Government Mining Journal* 70 (1969): 341; and D.C. Mengel, "Ipswich Coal-Field, Queensland", *Economic Geology of Australia and Papua New Guinea*, Vol.2, ed. D.M. Traves and D. King (Melbourne: Australasian Institute of Mining and Metallurgy, 1975), p.290.
23. The following articles claim that Logan first discovered coal on the Bremer River: R.L. Crivelli, "History of the Coal Industry in Australia", *Journal of the Australian Historical Society* 26 (1930): 132; Anon., "Ipswich—Pioneer Town of the Coal Industry", *Queensland Government Mining Journal* 78 (1977): 560; and H. Holthouse, *Illustrated History of Queensland* (Adelaide: Rigby Press, 1978), p.31.
24. Anon., "Ipswich-Pioneer Town", p.560.
25. C. Bateson, *Patrick Logan* (Sydney: Ure Smith, 1966).
26. "Journal of Captain Logan's Pedestrian Tour in the Neighbourhood of Moreton Bay in the Month of June 1827", *Sydney Gazette*, 17 August 1827.
27. Bateson, *Patrick Logan*; and L.R. Cranfield, "The Life of Patrick Logan", *Journal Royal Queensland Historical Society* 6 (1959–61): 302.
28. Russell, *The Genesis of Queensland*, p.145.
29. Meston, *Geographic History of Queensland*; and J.J. Knight, *In the Early Days* (Brisbane: Sapsford & Co., 1895), p.64.
30. Steele, *Explorers of Moreton Bay*, p.233.
31. A. Raistrick, *Industrial Archaeology* (London: Paladin, 1973), p.70.

Coal Use in the Mid Nineteenth Century

2

The closure of the penal settlement at Moreton Bay in 1842 permitted free access, travel and work within the previously-closed fifty mile radius of Brisbane Town but this did not in itself provide much of an incentive for the mining of coal from the substantial deposits which were known to exist in the region.

The only widespread use for coal in the early nineteenth century was as a fuel, for which duty it shows a considerable advantage over wood in terms of the amount of energy which it contains. Comparative tests[1] have indicated that bloodwood was probably the most effective wood fuel available at that time, but its specific energy (that is the amount of heat stored in a given weight of the material) was only about one half that of locally-occurring coal. Because wood is less dense than coal the total heat generated from a given volume of the best local timber would thus be barely a third of that emitted by the same volume of coal.

A high specific energy in a solid fuel is only of real importance where space is at a premium, such as in a steamship or a railway locomotive. For other heat-raising duties in a simple society, such as lime burning or brick making, wood is equally effective and was at that time more accessible than coal in most parts of northern New South Wales. Fraser, in 1828, had commented on the vast stands of excellent timber which lined the Brisbane River,[2] and during the years preceding Separation it is unlikely that the local population made significant inroads into them. Even in Sydney as late as 1833 it was unusual to burn coal domestically,[3] while wood was the sole source of fuel for locomotives in New South Wales until

1861.[4] The replacement of men by machines for heavy manual work which might have fostered an increased use of coal as a steam-raising fuel was in its infancy. For example, timber was cut entirely by hand in the Moreton Bay area until 1853 in which year William Pettigrew opened the first power-driven sawmill in Brisbane.[5]

At Separation in 1859 there were no more than four steam sawmills in the whole of Queensland while the total number of works and plants amounted to less than ten.[6] They concentrated chiefly on animal products with boiling down as the major manufacturing industry. The economy of the so-called Northern Districts of New South Wales which corresponded roughly with the present region of south Queensland was essentially pre-Industrial Revolution in character relying on wind, water or animals for power, and on timber for heat.

On the other hand the immigrant population was probably more used to burning coal for fuel than wood. Great Britain had made the transition from a wood to a coal-fired society during the century preceding 1750 because of the depletion of its forests;[7] the great majority of the immigrants reaching Australia would therefore have been thoroughly accustomed to using coal. The early explorers had kept a sharp lookout for it and there was no lack of knowledge regarding its plentifulness in the Moreton Bay area. But clearly its occurrence was not as widespread as wood and the population tended to use the fuel which was most immediately to hand.

The 1840s were years of economic crisis and recession in Australia reflecting the depressed conditions prevailing in Europe. The rapid advance of the pastoral industry and its growing importance as a supplier of raw wool to English manufacturers had turned sour and no alternative overseas market of comparable size could be found. The average price of wool exported from New South Wales in 1836 had been 2s. per lb (44 cents per kg at 1980 rates of conversion). In the following years it dropped steadily reaching 1s.1d. per lb in 1843 and falling to less than 1s. per lb by 1849. Land sales slumped, retail prices and wages declined by more than a third as compared with 1840 and by 1844 bankruptcy was widespread.[8] Despite these problems the volume of wool exported to the United Kingdom rose rapidly from £12 million ($24m) in 1841 to £39 million in 1850, raising the total value of the clip and calling for a greatly increased amount of transport from the pastoral regions to the coast. Most of the wool produced in the Northern Districts was shipped to Sydney from the growing port of Brisbane until the year 1851,[9] when direct overseas dispatch commenced with the sailing of the *Rebecca* for London. In 1842 Brisbane was handling an average of one hundred and fifty bales of wool per month.[10] This quantity had doubled by 1847[11] and reached one thousand bales per month by 1859.[12] The settlement of the Darling Downs and the Tablelands in the 1840s promoted Ipswich to a key position as the closest river terminal through which wool could be shipped to Brisbane by steamer or barge for reshipment, and made it the focal point for the distribution of incoming supplies. The river trade enjoyed an era of activity and enterprise which only died after the completion of the railway link between Brisbane and Ipswich in 1874.[13] In so far as the expansion of the wool trade called for additional steamer services on the Brisbane River and along the coast it encouraged the development of a coal industry to provide the fuel for the ships.

Plate 4. View of Brisbane, Moreton Bay, 12 November 1851 by Conrad Martens. (Courtesy Mitchell Library, Sydney.)

A secondary result of the low wool prices was the establishment of a substantial industry which concentrated on boiling down sheep and cattle for their fat and by-products. Dr Lang found in 1844 that "the second of the animal products of New South Wales was tallow".[14] In the same year the value of tallow exports from New South Wales was greater than the value of the entire wool clip of twelve years before.[15] But this was only the beginning. Boiling-down factories were started in Brisbane in 1843[16] and in Ipswich in 1847[17] and tallow exports from Brisbane rose from 296 casks in 1844 to 968 in 1847.[18] These activities certainly called for additional fuel but there was no real necessity for coal to provide it,[19] wood being equally effective[20] in most cases. Furthermore the natural oils and fats produced as by-products from the factories were readily available to the local population for lighting and heating, retarding the mining of coal for gas manufacture. The first coal-gas installation in Brisbane was not completed until 1857[21] and the floating of public gas utilities in Brisbane and Ipswich did not develop beyond the discussion stage until the mid 1860s.

Moreover in the 1850s when the price of wool was rising steadily and the economic scene was beginning to brighten in Australia, the Northern Districts of New South Wales could not expect to increase their fuel usage as a result of attracting a greatly increased population. The reason was that the influx of immigrants which followed the first discovery of gold in 1851 was confined to the south of Australia. It was not until 1858 that a find at Canoona near Rockhampton[22] brought a population surge to the north. In Victoria in the four years between 1850 and 1854 the gold rushes more than trebled its population, increasing its revenues and its produce sales twelvefold and raising its purchases twenty-four times.[23] In 1854 the population of Victoria reached a quarter of a million but it was a further four years before the embryonic colony of Queensland reached even a tenth of that figure. The manufacturing industries of Australia owed their origin to the exigencies of gold production, satisfying the need for near-at-hand facilities to meet the requirements of a vast new population.[24] In the Northern Districts there was steady pastoral development but the construction of a railway system, the production of gas for heat and light and the beginning of industrial production—forces that catalyzed coal production in the south of Australia in the 1850s—left Queensland almost untouched for at least another decade. Prior to Separation the total consumption of coal in the Northern Districts, other than in ships, seldom exceeded two thousand tonnes per year.

Ship Coal

The discovery of coal on the upper reaches of the Brisbane River by Lockyer in the earliest days of settlement,[25] and the reports of its occurrence by Cunningham in what is now the Ipswich area in 1828,[26] were the first of numerous sightings over wide areas of Queensland. On his 1844-45 expedition Leichhardt found coal at a number of places[27] between the Darling Downs and the Mackenzie River, while Clarke in 1852 estimated that the "Coalfield of the Condamine" on the Darling Downs covered an area of not less than fifty thousand square kilometres and he remarked on its cannel-like nature.[28] But clearly there was no point in mining any of this coal until a market could be found in close proximity to an easily-worked deposit.

The early mining of coal in Queensland is closely tied to the advent of the steamship and it is necessary to appreciate something of its development from a mechanical toy to a major factor in trade and economic relations between nations in order to understand the ebb and flow of coal mining in the Moreton Bay area in the middle of the nineteenth century.

Robert Fulton, an American, was the first to show the practical possibilities of steam navigation. In 1807 his *Clermont* made a journey of one hundred and fifty miles up the Hudson River in the remarkably short time of thirty-two hours. But it took almost a century for the steamship to triumph over the sailing ship. Unlike the railway, the steamer was not a new form of transport capable of serving places hitherto isolated from the main stream of commerce. Moreover, in the early years of the nineteenth century the Americans recognized the great possibilities of sail, and by the 1840s the clipper ship had evolved. Its great length, its fine lines and its huge expanse of canvas fitted it especially well for long voyages where speed and comfort were important considerations.[29] These conditions applied particularly between Europe, the United States of America and Australia.

Steamships first came into use on rivers, the sailing ship retaining its superiority on the high seas for many years after steamers were a common sight on river and coastal services, which they revolutionized by their speed and regularity. By 1843, Brisbane was connected with Sydney by a regular steamer service[30] and in less than a decade the sailing boats which were still left in coastal trading had been swept out of the passenger market.

In the case of intercontinental transport it was another matter. Heavy consumption of fuel and imperfections in boilers and engines gravely prejudiced the introduction of steamships on long voyages. Samuel Cunard successfully opened the first regular trans-Atlantic service in 1840[31] but the absence of coaling stations was a severe handicap on the long voyages from Europe to China and Australia. In 1847 less than 4 per cent of Britain's three million tonnes of shipping was propelled by steam,[32] and three years later over 95 per cent of shipping entering at or cleared from New South Wales ports was still under sail.[33] It was not until the second half of the century that the intercontinental steamer made rapid headway following the replacement of wood or iron by steel as the constructional material, and the invention of the compound engine in 1854.[34]

H.S. Russell[35] gives the date of the launching of the first steamboat in Australian waters—the *Surprise*—as 31 March 1831. Two months later, and before the *Surprise* could be completed, Lieutenant Biddulph sailed the redoubtable *Sophia Jane* into Sydney Cove and launched the first steamer service in Australian waters, operating between Morpeth on the Hunter River and Sydney.[36] This development quickly generated great optimism about the future of steamships, particularly as interest in regular trading with Asia via Cape York was quickening. In 1838 Port Essington—a fine natural harbour in Arnhem Land—was established and garrisoned as a routeing station, and the idea gained currency that profitable trading routes with Asia could and should be opened up.[37]

The depression of the early 1840s retarded all Australian shipping development but it did not stifle the ambitions of the residents of Moreton Bay to become a regular stop on international routes. In 1846 Moreton Bay became a port of entry to New South Wales[38] and there was great local indignation that Brisbane was not mentioned as a coaling station in the prospectus for a proposed steamship line for carrying mails from Europe between Singapore and Sydney via the Torres Strait.[39] Two and a half years later the *Moreton Bay Courier* was advocating the formation of a coal company to service coastal and international shipping with a regular, consistent supply,[40] but the proponents of the northern route were dealt a severe blow in 1849 with the abandonment of Port Essington.[41] The local paper continued to champion the cause[42] of a northern route and the New South Wales government persevered with an offer of financial support, although hedging its bets by offering a smaller subsidy for the establishment of a southern route to Singapore via West Australia.[43]

The death knell of the northern route was the acceptance in 1853 by the Peninsular and Oriental Company of the British government's contract to carry Australian mails from Singapore to Sydney via West Australian ports, Adelaide and Melbourne.[44] The decision was inevitable, bearing in mind the rapid growth that was occurring in the south of Australia, following the gold discoveries in Victoria and New South Wales. Although the New South Wales government continued to press the case for the northern route,[45] it was finally clear that Brisbane would not become a port of call on regular shipping services to Europe or Asia and would for the immediate future have to rely on river and coastal steamer traffic for markets for its coal.

Whether the introduction of a northern route would have established Brisbane as a major coaling port is, in any case, questionable. The hard fact had to be faced that despite the relative ease with which the coal could be mined on the banks of the Brisbane and Bremer rivers and loaded onto punts or similar vessels,[46] the river channel below Brisbane was too shallow to accept international steamers[47] and there was little economic or geographical justification for building coal-loading facilities on Moreton Bay itself. Entrance to the bay for most of the 1840s was gained through what later became known as the Southern Passage, a risky, shallow channel between the southern tip of Moreton Island and Amity Point on the northern end of Stradbroke Island. The loss of the *Sovereign* in the passage in 1847[48] spurred the government to search for a less dangerous entrance and Captain Freeman, who had been appointed Brisbane's first harbour master, brought into use the channel round the northern tip of Moreton Island which was later given his name.[49]

The bar at the mouth of the Brisbane River was notorious and reduced the depth of water to less than one and a half metres at low tide.[50] It was not until 1853 that any international steamship called at Moreton Bay,[51] and the first arrival was unable to take on coal at Brisbane because it could not cross the bar.[52] In 1851[53] the first of many river surveys was made but there was no progress in financing a dredging programme until the 1860s, by which time it was much too late to establish Brisbane as Australia's premier coal port. In 1847 the monopoly which the Australian Agricultural Company had held over coal production for the previous nineteen years in New South Wales[54] had been terminated by mutual agreement.[55] A surge of activity followed which doubled the coal production of the Newcastle area between 1849 and 1853 without satisfying the rising demand.[56] The high quality of the coal and the stagnation of Queensland's production during this vital period established Newcastle in an unassailable position as the principal coal port on the eastern Australian coast. When coal became more freely available in Brisbane in the late 1850s a small coastal trade in coal was finally established,[57] but the expansion of Brisbane into an international port which followed the commencement of the direct exporting of wool in 1851,[58] and cotton three years later[59] relied on the sailing ships whose need for coal was very minor and chiefly restricted to fuel for the galley.[60] As the mineralogical surveyor of the Australian Agricultural Company reported in 1852, "The coal seams at Moreton Bay will remain of very local interest and their produce would merely supply the few steamers to that harbour, but will not appear on the Sydney or Melbourne Markets".[61] Brisbane had "missed the boat" as an international coal port.

Coastal Activities

Following the arrival of the *Sophia Jane* in 1831, the use of steamships for coastal transport steadily expanded. Six years later the *James Watt* under charter to the New South Wales government became the first steam vessel to enter Moreton Bay, carrying amongst its passengers Andrew Petrie and his family.[62] In 1839 the Hunter River Steam Navigation Company was established[63] and it inaugurated a service between Sydney and Brisbane in late 1841.[64] It began at a time when the depression was deepening and after five months of unfavourable returns the company withdrew from the trade, but three months later it came back determined to prove that the first decision was correct. By April 1843 the economic conditions were so bad that the *Sydney Morning Herald* recorded that nine steamships were laid up in the port[65] but the company persevered with its new route and finally won through.

The coastal ships were small in comparison with the 800 to 1,000 tonnes displacement of the high seas vessels. The *Shamrock* which opened the fortnightly service[66] displaced only 320 tonnes,[67] its hull being approximately the same length as the body of a modern interstate jet liner and about half as wide again (plate 5). For the three or four

Plate 5. The steamship *Shamrock*. (Courtesy John Oxley Library.)

days normally required for the voyage from Sydney to Brisbane up to 160 souls were crammed into this small space, many passengers having no berth or place to lie down. Lifeboats were provided for barely half the complement and the ship was frequently grossly overloaded with wool and other animal products, especially on the return trip to Sydney.[68] The efficiencies of the boilers and the low-pressure side-lever engines were poor and the consumption of fuel was high. In 1828 a contemporary British specialist, Sir John Ross, established the coal requirements of a marine engine by the following procedure;

> . . . one bushel [of coal] is the allowance usually made for each horse power for a day's work; but the day when spoken of as applying to horse's work, is but 8 hours of actual labour. This allowance has by many been considered as insufficient, and for small engines it is perhaps not quite ample. But there is no doubt, that if an engine is kept constantly worked, one

> bushel per horse power for each 8 hours work will be fully sufficient.[69]

Assuming that a bushel of coal weighs 80 lbs (36 kg), the 100 horsepower (75 kW) engine of the *Shamrock* consumed about 10 ½ tonnes per day, a figure which is confirmed by contemporary records.[70] The bunker capacity was 86 tonnes which was sufficient for a round trip although the vessel occasionally took considerably more on board with the excess being stored on deck, to the discomfort and inconvenience of the passengers.[71] The *Sovereign*, which took over the service in 1843[72] and made the trip regularly until wrecked in dramatic circumstances while fighting her way through the Moreton Bay South Passage in 1847,[73] was even smaller than the *Shamrock*, displacing only 214 tonnes.[74]

The Hunter River Steam Navigation Company preferred to burn coal rather than wood even when it was more expensive, because of the small size, low efficiency and high consumption of their ships.The coal had to be purchased in Sydney for the round trip[75] at the price demanded by the monopoly suppliers, the Australian Agricultural Company. In 1843 the rate was 17s. per ton ($1.70 per tonne)[76] and the steamship company threatened to terminate the Brisbane service unless local bunkering facilities could be provided.[77] Their Brisbane agent, Mr Connelly, therefore approached a certain John Williams and induced him to search for coal.[78] After he had found a source at Redbank and government approval had been received for working it, a contract was signed with the manager of the company, Francis Clarke, for the regular coaling of their ships in Brisbane.[79] The story of the development of Williams's coal interests on the Brisbane River is recounted in a later chapter.

One result of the contract is perfectly clear—it did not provide the steady, consistent supply of coal which the steamers to Sydney required. To what extent the frequent criticisms by the public of the company (and after 1851 of its successor the Australasian Steam Navigation Company) can be blamed on inadequate fuel supplies is difficult to determine but they cannot have been inconsiderable. Their ships were sometimes so short of coal that they could not sail for Sydney[80] or they had to resort to burning the wood which was stacked on deck for this purpose. When the *Sovereign* foundered in March 1847 while leaving Moreton Bay en route for Sydney, the injuries which some passengers sustained were received from the billets of wood carried on deck for fuel which broke loose as the vessel struck.[81]

The removal of the coal monopoly in 1847 benefitted the Newcastle coal mines much more than those in the Northern Districts and the first real chance for the Moreton Bay suppliers to make a challenge for the coastal coal trade came with the discovery of gold in Bathurst in 1851. A scarcity of labour of all kinds quickly developed in the south of New South Wales as every able-bodied male with a horse of his own, or a pair of legs to move him and shoulders to support his bundle, set out post-haste from the coastal plains over the Blue Mountains for Ophir or other goldfields in the area. Those who stayed behind had the labour market at their mercy for a few years. The records of the time are strident with the complaints of the employers about scarcity of workers and the insolent independence of those whom gold had lured from their normal occupations. The coal miners remaining in Newcastle took full advantage of their position and the price of coal at the staithes rose from 6s. 6d. per ton (65 cents per tonne) in 1850 to 20s. 2d. in 1854.[82]

In this period the Australasian Steam Navigation Company came to rely quite heavily on Brisbane for recoaling its ships because the price rose much less steeply than at Newcastle.[83] But the company also exploited the gold rush trade by putting every available ship on the Sydney to Melbourne line and reducing the frequency and regularity of the Brisbane service.[84] This had one useful result in that it stimulated the direct shipment of wool to Europe from Brisbane, although it made little difference to the coal market because the service relied solely on sailing ships.

The Brisbane citizens felt so strongly about their neglect by the Australasian Steam Navigation Company that in April 1852 they attempted to start their own shipping line.[85] This was as unsuccessful as the attempts of several independent operators but the warning was clear, and in 1854 the shipping company arranged to reopen the regular Brisbane service with the *Shamrock* and the newer, screw-driven, *Boomerang*.[86]

Not only were the Moreton Bay mines unable to take advantage of the steep rise in the price of coal which occurred between 1851 and 1854,[87] but the largest producer, Moggill Mine, could not even meet local commitments. In 1852 it was closed by floods[88] and in December 1853;

> Mr. Williams' coal pit on the Brisbane River fell in . . . owing to the workmen having, in technical parlance 'robbed the pit', that is made use of the pillars of coal that supported the roof. Providentially there was no injury to life or limb, but as Brisbane only possesses one man with sufficient enterprise to work the coal fields, this accident will stop the supply of coal until a fresh opening can be made, which will not be under a month hence.[89]

This estimate was unduly optimistic and by early 1854 the supply position in Brisbane was so bad that one of the steamers to Sydney was unable to sail for lack of coal.[90] Nor were conditions improving in Newcastle, where in

May 1854 the miners struck for a further rise in the hewing rates.[91] Over a period of a few months there was a flurry of proposals for taking advantage of the position by starting new mining operations on the Brisbane and Bremer rivers[92] but most of them came to nothing.

The editor of the *Moreton Bay Courier* reflected the frustrations of the time in a leading article on developing steam navigation and coal mining in the Moreton Bay area.[93]

> The People have so long been under the control of timid doubts and fears, that much time will have elapsed before they can be made fully to appreciate their own advantages, and the undeveloped wealth of the country around them. As it is sail [sic] that the horse is kept in subjection by the man for fear that he should discover his own strength and shake off the tyrannic controul [sic] of his puny master, so the people of Moreton Bay have endured a condition of subserviency until they have lost confidence in themselves, and therefore find difficulty in securing the confidence of others. Harsh words these but true. It is to England therefore that we must look for the energy and the capital necessary to work the mines of the Northern Districts: and those who might have reaped the golden harvest themselves, must stand aside and make way for the courageous and far-seeing English capitalist.

In Newcastle the final crisis was reached in 1855 with a new stoppage by the miners which was intended to force the coal owners to reinstate the provision of free tools and light, traditionally supplied by the mines. This demand met with firm resistance, the men having chosen an unsuitable time to apply pressure. Starving gold miners from the western fields were beginning to return to their old environment in woebegone processions ready to take any job for any pay. The price of coal fell quickly and before the end of the decade it was down to 14s. ($1.40) per tonne.[94] The steamship company abandoned Brisbane as a coaling station, finding it cheaper and more reliable for its ships to make a two hour stop at Newcastle on their northward run from Sydney.[95] It was not until the gold rush to the Fitzroy River in 1858 provided the company with a more northerly terminal port and Redbank began selectively mining a better quality product[96] that Brisbane once again began to increase its coal sales to coastal steamers. The quantity of coal consumed by the coastal vessels was not large, the statistics showing that the total amount leaving Brisbane or Moreton Bay as bunker fuel or cargo seldom exceeded one thousand tonnes per annum or an average of twenty tonnes per week.

River Traffic

Steamer traffic on the Brisbane River commenced in 1846 when the *Experiment*, a converted horse-operated boat, was put into service between Brisbane and Ipswich. It was an immediate success, and despite running aground on its maiden voyage and subsequently being sunk at its moorings it gave good service until finally broken up in 1849.[97]

The river steamers were much smaller than the coastal packets, possessing a very shallow draught. This was essential because of the state of the river between Brisbane and Ipswich which at low tide had a depth of

Plate 6. Portion of Conrad Marten's sketch "Ipswich Landing Place", dated 24 November 1851. (Courtesy Mitchell Library, Sydney.)

only three quarters of a metre in some places.[98] The *Experiment* drew a mere 0.6 metre of water and displaced thirty-seven tonnes.[99] Shortly before she was broken up in 1849 the Hunter River Steam Navigation Company introduced the larger *Raven* on the run for a short period but her sailing times had to be determined by the tides because of her draught and she did not stay long in service.[100] The locally-built *Hawk*,[101] which ran for most of the 1850s in collaboration—or competition—with half a dozen other steam paddle boats, had a draught of less than a metre.[102] Their engines were quite small, usually between twelve and twenty-four kilowatts, and there was seldom more than one boat per day scheduled to make the journey between Brisbane and Ipswich (plate 6). It was reasonably convenient for these steamers to take on coal from the mines situated on the banks of the Brisbane River although this was obviously not as satisfactory as re-coaling them at Ipswich or Brisbane while they were being turned round. An attempt was made to satisfy this need in 1851 when Reid and Boyland opened a small mine on the Bremer River[103] at Ipswich, but the operation was short-lived, probably because of the poor quality of the fuel. A couple of years later Thomas Boyland worked a small mine on his land in Ipswich to fuel the *Hawk* which he owned at the time.[104] In the late 1850s coal supplies became more readily available from Walter Gray's new mine at Ipswich and particularly from John Campbell's Redbank operation[105] so that the pressure to open other pits to fuel the river steamers disappeared.

The total consumption of these river steamboats probably added up to about the same as that of the packet boats running to Sydney; it is extremely unlikely that their total consumption ever exceeded a thousand tonnes per annum.

In summary, then, the total requirements for coal in the Northern Districts of New South Wales during the 1840s and 1850s were quite small, seldom in excess of five thousand tonnes per annum. Much of this was required for bunker fuel but there was a growing demand for export and from local industry, including the small number of boiling-down plants, forges and sawmills in the district.

Notes and References for Chapter 2

1. W. Fryar, Annual Report to Department of Mines, 1899 (Brisbane: Queensland Government Printer), pp.156-64.
2. J.G. Steele, *The Explorers of the Moreton Bay District, 1770-1830* (St. Lucia: University of Queensland Press, 1972), p.233.
3. *Sydney Gazette*, 30 May 1833.
4. G.A. Gelder, "Early History of the Railways of New South Wales", *Royal Australian Historical Society* 17 (1931): 215.
5. W. Pettigrew, Personal Papers. Royal Historical Society of Queensland, Newstead House, Brisbane.
6. *Statistics of Queensland*, 1860.
7. H. Hamilton, *History of the Homeland* (London: George Allen & Unwin Ltd., 1947), p.163.
8. B.Fitzpatrick, *The British Empire in Australia* (Carlton: Melbourne University Press, 1941), pp.109-13.
9. *Pugh's Almanac* (Brisbane: T.P. Pugh, 1859), p.87.
10. T.A. Coghlan, *Progress of Australia in the 19th Century*, Vol. 1 (Sydney: Government Printers, 1903), p.299.
11. J.J. Knight, *In the Early Days* (Brisbane: Sapsford & Co., 1895), p.247.

12. M. Fox, *The History of Queensland*, Vol.1, Part 2 (Brisbane: State Publishing Co., 1919), p.440.
13. *Votes and Proceedings of the Legislative Assembly of Queensland*, 1876, p.567.
14. Fitzpatrick, *The British Empire in Australia*, pp.109–13.
15. Ibid.
16. *Pugh's Almanac*, 1859, p.87.
17. *Moreton Bay Courier*, 9 January 1847.
18. Knight, *In the Early Days*, p.247.
19. *Moreton Bay Courier*, 4 October 1851.
20. Ibid., 20 March 1847.
21. Ibid., 25 July 1857.
22. Ibid., 9 October 1858.
23. Fitzpatrick, *The British Empire in Australia*, p.160.
24. Ibid., p. 249.
25. E. Lockyer, "Journal of an Excursion up the River Brisbane in the Year 1825", *Australian Quarterly Journal* 1, ed. C.P.N. Wilton. See also H.S. Russell, *The Genesis of Queensland*, Appendix C, (Sydney: Turner and Henderson, 1888), p. 589.
26. H.S. Russell, *The Genesis of Queensland* (Sydney: Turner and Henderson, 1888), p.145.
27. L. Leichhardt, *Journal of an Overland Expedition in Australia* (London: P. & W. Boone, 1847).
28. *Votes and Proceedings of the Legislative Council of New South Wales*, 1853, Vol.2, pp.853–70.
29. Hamilton, *History of the Homeland*, p.350.
30. *Sydney Morning Herald*, 14 September 1842; and N.L. McKellar, *From Derby round to Burketown* (St. Lucia: University of Queensland Press, 1977), p.7.
31. Hamilton, *History of the Homeland*, p.352.
32. Ibid.
33. M.H. Ellis, *A Saga of Coal* (Sydney: Angus & Robertson, 1969), p.172.
34. Hamilton, *History of the Homeland*, p.352.
35. Russell, *The Genesis of Queensland*, p.64.
36. Ellis, *A Saga of Coal*, p.28.
37. G.C. Bolton, *A Thousand Miles Away* (Canberra: Australian National University Press, 1970), p.9.
38. Knight, *In the Early Days*, p.356.
39. *Moreton Bay Courier*, 5 December 1846.
40. Ibid., 26 February 1848.
41. Bolton, *A Thousand Miles Away*, p.11.
42. *Moreton Bay Courier*, 4 October 1851; 2 October 1852; and 3 December 1853.
43. *Votes and Proceedings of the Legislative Council of New South Wales*, 1850, Vol.2, pp.445–47.
44. Fitzpatrick, *The British Empire in Australia*, p.106.
45. *Votes and Proceedings of the Legislative Council of New South Wales*, 1855, Vol.2, p.1027.
46. *Moreton Bay Courier*, 22 April 1848.
47. Ibid., 23 December 1848.
48. Ibid., Extraordinary Issue, 17 March 1847.
49. McKellar, *Derby to Burketown*, p.8.
50. E.V. Stevens, "Port of Brisbane", *Journal Historical Society of Queensland* 4 (1952): 667.
51. *Moreton Bay Courier*, 7 May 1853.
52. Ibid., 14 May 1853.
53. Ibid., 5 July 1851.
54. J. Gregson, *The Australian Agricultural Company, 1824–1875* (Sydney: Angus & Robertson, 1907), p.50.
55. Earl Grey to Sir Charles Fitzroy. Despatch 127, 13 April 1847.*Historical Records of Australia, Series I*, Vol.25, pp.486–88.
56. H.W.H. Huntington, "Our Northern Coal-Fields", *Sydney Quar-*

terly Magazine (1889): 256–68; and *Moreton Bay Courier*, 10 September 1853.
57. *Moreton Bay Courier*, 15 June 1859.
58. Ibid., 14 September 1850.
59. Ibid., 10 June 1854.
60. W. Johnston, "Old Bulimba", *Journal Historical Society of Queensland* 1 (1918): 304–18.
61. Papers of the Australian Agricultural Company. New South Wales Archives, Location 2/2322.
62. C.C. Petrie, *Tom Petrie's Reminiscences of Early Queensland* (Brisbane: Watson Ferguson, 1904), p.257.
63. McKellar, *Derby to Burketown*, p.5.
64. N.S. Pixley, "The History of the A.U.S.N. Co. Ltd. and its Predecessors", *Journal Historical Society of Queensland* 5 (1954): 969–1003.
65. *Sydney Morning Herald*, 20 April 1843.
66. Ibid., 2 November 1842.
67. *Votes and Proceedings of the Legislative Council of New South Wales*, 1854, Vol.2, pp.821–28.
68. Ibid., 1852, Vol.2, pp.1007–39.
69. J. Ross, *A Treatise on Navigation with Steam* (London: Longman, Rees, Orme, Brown & Green, 1828), p.168.
70. *Votes and Proceedings of the Legislative Council of New South Wales*, 1852, Vol.2, pp.1007–39.
71. Ibid.
72. McKellar, *Derby to Burketown*, p.13.
73. *Moreton Bay Courier*, Extraordinary Issue, 17 March 1847.
74. A.B.Portus, "Early Australian Steamers", *Journal Australian Historical Society* 2 (1904): 197–226.
75. Pixley, *History of the A.U.S.N. Co.*,pp.969–1003.
76. *Votes and Proceedings of the Legislative Council of New South Wales*, 1847, Vol.2, pp. 335–92.
77. A.K. Thomson, *The Collected Works of Thomas Wesley*, Vol.1 (Brisbane: Jacaranda Press, 1967), p.302.
78. Johnston, "Old Bulimba" (note by W. Clark), p.314.
79. *Moreton Bay Courier*, 26 July 1851.
80. Ibid., 26 February 1848.
81. Ibid., Extraordinary Issue, 17 March 1847.
82. Ellis, *A Saga of Coal*, p.42.
83. *Moreton Bay Courier*, 8 October 1853.
84. McKellar, *Derby to Burketown*, p.19.
85. *Moreton Bay Courier*, 24 April 1852.
86. *Votes and Proceedings of the Legislative Council of New South Wales*, 1852, Vol.2, pp.1007–39; and *Moreton Bay Courier*, 4 March 1854.
87. *Moreton Bay Courier*, 8 October 1853.
88. Ibid., 17 April 1852.
89. Ibid., 17 December 1853.
90. Ibid., 4 March 1854.
91. Ellis, *A Saga of Coal*, p.41.
92. *Moreton Bay Courier*, 10 September 1853; 11 March 1854; 6 May 1854; 27 May 1854; 24 June 1854; and 20 October 1855.
93. Ibid., 1 October 1853.
94. Ibid., Extraordinary Issue, 17 March 1847.
95. Ibid., 29 September 1855; 20 October 1855; and 3 November 1855.
96. Ibid., 9 October 1858; and 13 March 1860.
97. Ibid., 13 June 1846; 20 June 1846; 22 January 1848; and 29 September 1849.
98. *Votes and Proceedings of the Legislative Council of New South Wales*, 1855, Vol.2, pp.1117–22.
99. H. Norrie, "Sydney's Ferry Boats", *Royal Australian Historical Society* 21 (1936): 1.
100. *Moreton Bay Courier*, 2 September 1848; and 2 March 1850.

101. Ibid., 2 March 1850.
102. L.E. Slaughter, *Ipswich Municipal Centenary* (Ipswich: City Council, 1960), p.41.
103. *Moreton Bay Courier*, 23 April 1853.
104. Ibid.
105. *Statistics of New South Wales*, 1858, p.84.

The Early Mining of Coal

3

Following Cunningham's reference to the existence of coal on the banks of the Bremer and Brisbane rivers in 1828 there is a long silence about its being mined, although Meston[1] claimed that coal was used in small quantities during part of the convict period from 1828 to 1839. Pugh, in his first almanac published in 1859,[2] ignored Cunningham's reports and maps and attributed the discovery of coal on the Brisbane and Bremer rivers to Andrew Petrie, believing that the discovery occurred shortly after Petrie's arrival in 1837 in the *James Watt*. Pugh describes how, in 1838, Petrie sent a sample to Sydney where "... it was pronounced to be of excellent quality by competent judges".[3] Knight is even more specific and attributes to Andrew Petrie the discoveries at Redbank and Moggill.[4] He refers to Petrie's 1838 sample as being discovered at Tivoli on the Bremer River and describes how "... later a tunnel was run into the hill and a plentiful supply obtained for the penal settlement". Tom Petrie's references to the workings of coal during the convict settlement period appear to have been taken directly from Knight.[5] The basis for these various legends is probably to be found in a letter which was despatched from the Moreton Bay settlement to the colonial secretary's office in Sydney in November 1837.

At the time the commandant of the penal settlement at Moreton Bay was Major Cotton and in his regular report to the colonial secretary's office he included the following significant paragraphs:

> I feel it to be my duty to bring under the consideration of his Excellency the Governor the Section of a seam of Coals, which

> are found on the banks of the "Brisbane" about two miles from the Limestone Station and fifty from Brisbane Town.
>
> Shortly after the arrival of the 'James Watt' steamer in Moreton Bay, I inspected this Coal and afterwards requested Mr. Petrie to do the same and to furnish me with a report upon it.
>
> By the first such opportunity I have forwarded to the care of the Master Attendant in Sydney samples of the Coal, of which and of the appearance of the Mine Mr. Petrie thinks very favourably and I feel confident that a supply of Coals of the best quality might be held in leadings [?] at a depôt in the bay, for the use of Steamers, if Government should send for timber by such conveyance.[6]

The reply, despatched from Sydney six months later asked for further details but made it clear that the exclusive privilege granted to the Australian Agricultural Company in 1828 for supplying coal to the public would prevent anything being done with respect to working the Moreton Bay seam without the permission of the secretary of state.[7] There the matter rested until the advent of a steamship service between Sydney and Brisbane four years later.

The Hunter River Steam Navigation Company turned its attention to providing a regular service to Moreton Bay in 1841, and in December of that year one of its newest boats, the *Shamrock*, made a trip to Brisbane in order to assess its practicability.[8] The cost of coal was a major item in the budget of the company[9] and there is no doubt that the possibility of establishing coaling facilities in Brisbane and thus breaking the monopoly privileges of the Australian Agricultural Company provided a secondary motive for the voyage. A precedent had been set in 1838 when limited production was permitted at Port Phillip because it was considered unreasonable to haul the coal seven hundred miles from Newcastle.[10]

The Australian Agricultural Company recognized the danger and took what steps it could to protect its position. The company's commissioner, Captain P.P. King, appraised the situation in a letter written to their head office in London in February 1842:

> . . . of future importance—is the Moreton Bay mines where I am told the Coal is abundant and of good quality. It is reported that upon the application of the Hunter River Co—to the Governor for permission to work these Coals—the Governor said he had no objection if I had not—. One of the Iron Boats went there last week and met with so much encouragement that I think she will go every month and if permitted to work she will embark as much coal as will take her to Sydney—say 200 Tons a month— This is not of very serious consequence but it will be a growing rival which should be stopped and Mr. Croasdill who is ever on the alert and fertile in all his schemes sharply recommends that the Company should detach a few men to work it so that the Governor would have no right to permit others to do it— To do

> this the allotment must be purchased and probably a large sum would be necessarily paid for it—and it might be many years before we could sell enough to make it an object of consequence—I shall communicate with the Government upon the subject and see what is best to be done to maintain your rights . . . Of Port Phillip I am not afraid—but of Moreton Bay I am afraid. I do think you ought to make some exertion to locate yourself there—of which the Establishment would not be a very expensive one as the Coal is close to the surface.[11]

While legal representations were being made by the company in London the Hunter River Steam Navigation Company wasted no time in making approaches to John Williams who, as mentioned in the previous chapter, immediately accepted the commission of attempting to provide coal for their steamships.

Williams was one of the first great entrepreneurial characters of early Queensland. He was a native of Somerset where, it was said,[12] he once worked as a subcontractor in one of the county's coal mines situated at Radstock. But in the light of Samuel Stutchbury's later strictures on his operations at Moggill[13] it is hard to believe that Williams's mining experience amounted to very much. What is clear is that he was a seaman and salesman of no mean skill, having been the first regular sea trader between Sydney and Brisbane[14] until 1841 when, at the age of forty-four,[15] he settled in Brisbane. Here he established the first public house,[16] became the first lessee of the Russell Street ferry[17] and continued to build, own and operate boats, punts and barges.[18] His first shaft, presumably for prospecting, is said to have been sunk at Fairfield,[19] and some coal was apparently found at Softstone on Oxley Creek[20] although it has not been possible to identify the place. Unfortunately neither of these prospects was satisfactory, yielding only thin beds of coal and masses of decayed fossils.[21] His attention therefore moved farther up the river to places where Andrew Petrie, the superintendent of works, had located coal some five years earlier[22] while selecting suitable quarrying areas.[23]

Coal outcropped on the south bank of the Brisbane River at Redbank some two hundred metres below Six Mile Creek, dipping away under shales, sandstone and conglomerates,[24] and had been visible in the vicinity of the track between Brisbane and Ipswich from an early period in the history of the settlement.[25] It was at this point on what was later to be known as Portion 129 Goodna that Williams opened his mine, supposedly called the Pioneer[26] (plate 7). The government legalized the operation by granting him a "right or privilege" to mine the coal[27] although the full terms of the grant are unfortunately unknown.

There are some discrepancies in the literature regarding the actual year in which production began, ranging from

1843 through 1844 to 1846.[28] It is most likely that mining began in the early months of 1843 to fuel the steamship *Sovereign* which was attempting to maintain a regular service between Sydney and Brisbane. The timing was unfortunate for Williams because the economic recession

Plate 7. The site of John Williams's first mine at Redbank, to the left of the bushes on the riverbank in the centre of the photograph. Morris's woollen mills may be seen in the background. (Photographed in June 1980.)

was at its deepest in 1843 and he apparently over-extended his tenuous financial resources when he moved into coal mining. At the end of the year he was forced to sell not only the Captain Piper Inn and the Brisbane ferry (for which he held the licence) but also ". . . the punt or barge plying between Brisbane and Redbank for coal, with all his interest in the coal now gotten or to be obtained by Mining for the same and the right or privilege appertaining thereto".[29] The sale realized only £150 ($300) and there is no evidence that the purchaser, John McConnel, had any interest in operating the mine at Redbank, although he probably continued to do so.

The persisting difficulties of the Hunter River Steam Navigation Company mentioned in the previous chapter suggest that the supply arrangements were in fact far from satisfactory. This was not helped by the shallowness of the river above Seventeen Mile Rocks,[30] the periodic floods which the Brisbane River was experiencing at that time,[31] or the geology of the seam at Redbank which dipped downwards from the river bank exposing the mine to flood waters[32] which could not drain away when the river subsided. Nor, judging from later reports on Redbank coal,[33] was its quality really adequate for the coastal packet boats. In 1846 the unsatisfactory coal supply position was further aggravated by the introduction of a regular steamer service between Brisbane and Ipswich. A reporter from the *Moreton Bay Courier* who took a trip on the *Experiment* from Brisbane to Ipswich in August 1846 mentions that the coaling stop was at the "Coal Reach" situated on the Brisbane River between Woogaroo Creek and the junction with the Bremer River.[34] This suggests that the mine at Redbank was still operating at that time.

Meanwhile the Crown persisted in upholding the rights of the Australian Agricultural Company and the secretary of state for the colonies continued to insist that he and not Governor Gipps should decide whether or not each particular case fell within the terms of the monopoly.[35] The uncertainty of the law governing the mining of coal, the limited demand for it and a steady fall in its price no doubt discouraged Williams from starting another venture.

In 1847, following an enquiry into the current procedures for mining coal, the monopoly privileges of the Australian Agricultural Company were swept away,[36] and in early 1848 government policy underwent a complete change. It was decided that "The interests of the Colony would be best served by affording the utmost encouragement to the raising of coal"[37] and the views expressed in 1840 by the secretary of state for the colonies of the day, Lord John Russell, that "All deeds of grant, throughout the whole of the Colony, should henceforward convey to the purchaser everything below and everything above the surface"[38] were applied retrospectively to coal although the official proclamation was not issued until 1850.[39]

The reversal in government policy in early 1848 came at a time when the coal supply position in Brisbane was particularly unsatisfactory. The *Moreton Bay Courier* summed up the position as follows:

> The formation of a Coal Company with a moderate capital has long been a desideratum in this district; for it is not to be supposed that steamers will visit this port unless they can readily obtain their supply of coal. The fact of the "Thistle" having been detained for this cause, and also the "Cornubia", has attracted our attention to the subject; and we are of the opinion that a company, with a capital of £1000, in forty shares of £25 each, might be formed at an early period. If managed economically, we have no doubt whatever that its success would be such as to cause the shares to rise to a premium after the preliminary expenses shall have been paid off;

and that its establish[ment] would be one of the greatest benefits that could be conferred on the district. The time is ripe for such an undertaking, and we hope to see it shortly carried out.[40]

This was exactly the encouragement that John Williams needed to re-enter the coal mining business and in 1848 he commenced operations on the opposite bank of the Brisbane River from Redbank, at Moggill, where there was an outcropping deposit of higher quality.[41] Moreover, when the area containing the deposit was subdivided in 1849, following instructions from the surveyor-general to provide "Special Country Lots" of less than forty hectares each,[42] Williams bought some twenty-four hectares knowing that under the new government policy he was also buying any coal which lay beneath them (document 2). He gained possession of Portion 14 Moggill on 8 April 1850 and of Portion 15 Moggill, lying immediately to the south of Portion 14 (and containing the mine itself), on 5 August 1851.[43] The mining potential of the blocks was recognized by the purchase price being increased from the standard upset price for grazing land of £1 per acre ($5 per hectare)[44] to £2 per acre for the sixty acres (twenty-four hectares) concerned.

Moggill Mine

Williams was a taciturn character[45] and none of his writings or records have so far come to light. We can, however, gain a good idea of Moggill Mine from two authors, Samuel Stutchbury and William Fryar. Stutchbury was an English geologist and biologist with an excellent reputation earned over many years of professional work. In 1844 he had advised the British Government on the Haswell Colliery explosion in County Durham where ninety-five miners lost their lives, and he had gained wide success as a coal viewer, which was a senior position in the management of a British colliery.[46] In 1850 at the age of fifty-two he applied for, and was appointed to, the position of geological surveyor to the New South Wales government (plate 8). In September of the following year[47] a memorial was submitted by Brisbane residents to the governor general praying that Stutchbury should visit the Moreton Bay area in order to assess its potential for gold mining. The government assented but when in 1853 Stutchbury finally arrived his official instructions were to give particular attention to coal " . . . for the purpose of ascertaining whether a royalty should be imposed" rather than to look for gold.[48] The most important visit was to be to Moggill Mine but this was unavoidably delayed by a fall in the roof which temporarily closed the workings[49] and he finally reached there early in 1854. He reported:

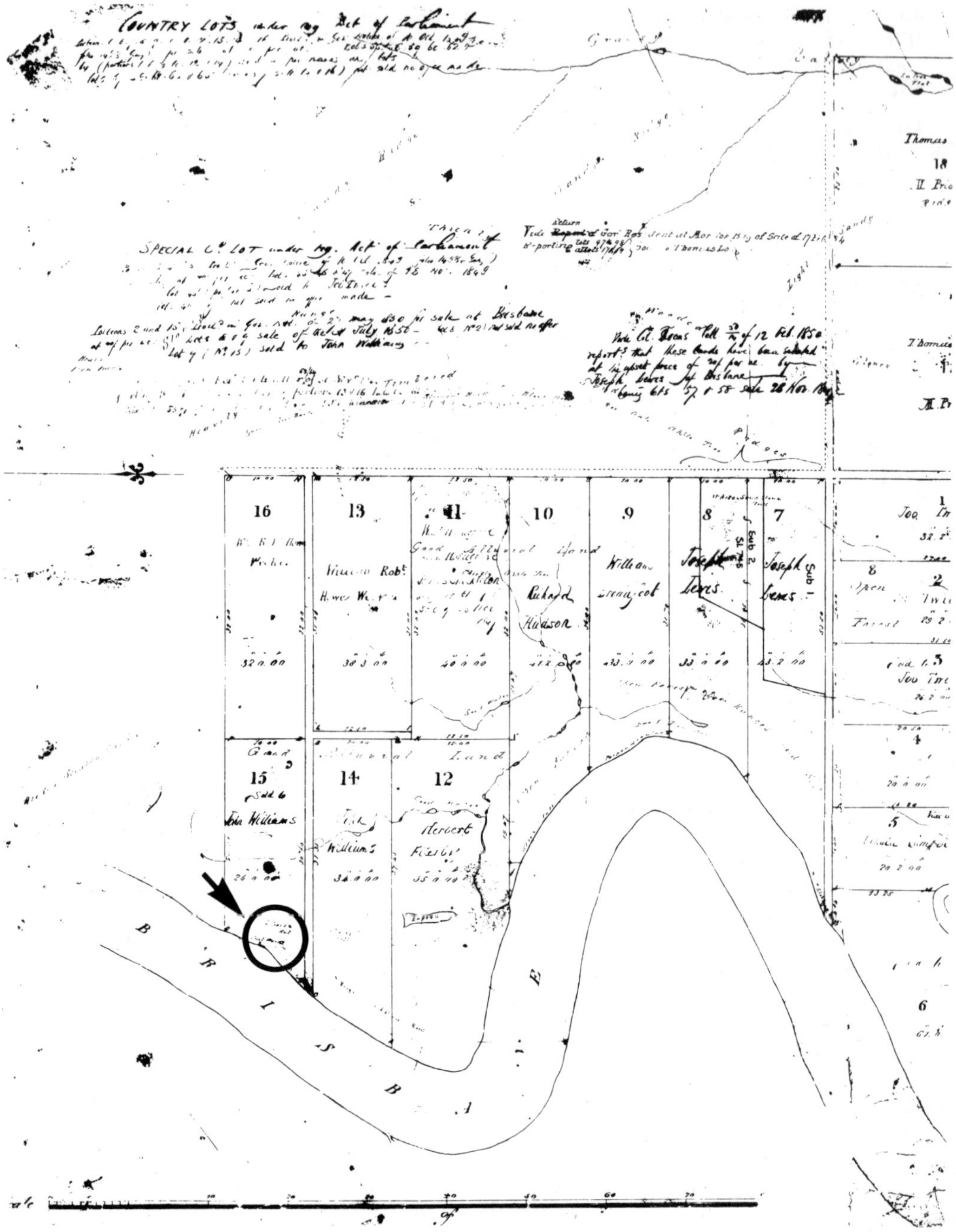

Document 2. Part of the 1849 subdivision plan of Moggill (chap. 3, n. 42) showing Williams's "coal works". (Plan supplied by the Surveyor-General, Queensland, and reproduced by arrangement with the Queensland Government. CROWN COPYRIGHT RESERVED.)

Having returned to Brisbane for the purpose of getting a new departure, I reached the first coal "crop", in the parish of Moggil [sic], on Mr. Williams's purchased land of sixty acres. Mr. Williams, as I have before mentioned, being the first person who has to any extent adventured in the coal trade. At this place I found the coal exposed on the river bank, near the water's edge.

Plate 8 *Samuel Stutchbury* (artist unknown). (Courtesy Mines Department, Queensland Government.)

The works have been commenced by a "day level"[a] running in from the bank, the principal "way" being six feet wide. The "top way side"[b] is not worked to any great extent, in consequence of the very thin covering of the superstrata. The "deep way"[c] workings are more extensive, being about sixty yards down from the head of the "horse way"[d].

a. Day level. Scottish term for a level or open cutting driven from the surface to the workings.
b. Top way side. Up-dip from the main way or principal passage.
c. Deep way or deep side. Down-dip from the main way or principal passage.
d. Horse way. An underground road fit for travel with a horse.

The coal varies from five to six feet in thickness, dipping south forty degrees east, angle from seven to eight degrees, although not worked its whole thickness, excepting where they wanted head-room for the "horse way". The main coal separates from the top coal at about four feet; the two feet of

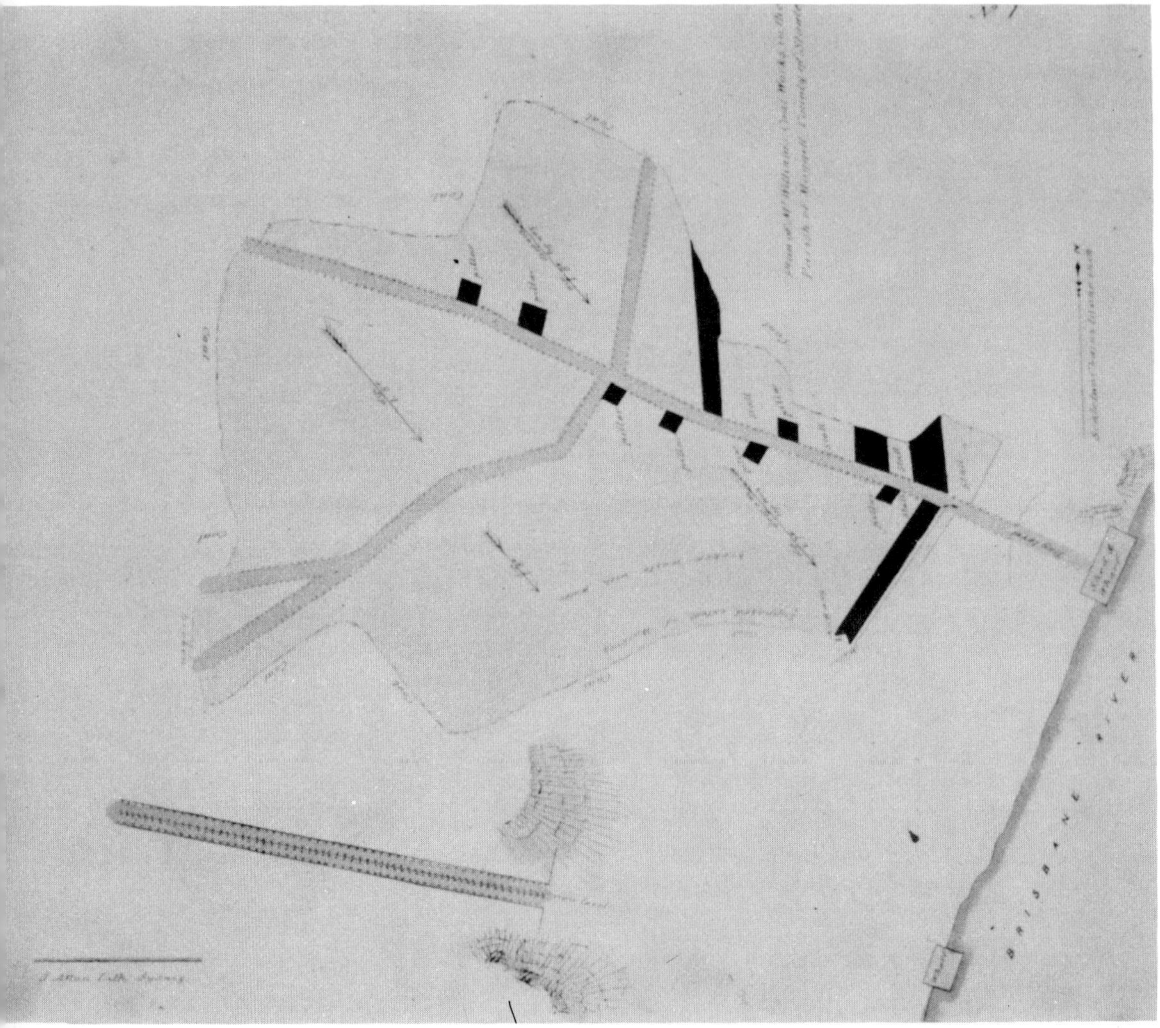

Document 3. Stutchbury's plan of Moggill Mine made in 1854 (chap. 3. n. 50), the earliest known plan of a mine in Queensland.

the top coal is good, but mixed with a small quantity of earth; it has not been worked, because the men would not take the trouble to "hand-pick" it. I therefore consider that the proprietor has lost at least one-third.

From the want of knowledge of coal working in master and men, I have never in my experience seen such want of care, and consequent waste; the neglect in proper timbering (inexcusable where timber is to be procured by the mere labour of cutting and fetching it) has rendered the whole of the "deep side" workings a broken waste, with much water.

The water that I saw was consequent upon the late rains, which had filtered down the cracks and fissures which passed upwards from the broken mine to the grass land; yet notwithstanding that they have been troubled with water, they have had the benefit of a free ventilation by means of the same fissures, the ventilation not having been provided for in any other way.

Much of the surface land is in progress of settling down into the "wastes" of the mine; in one instance the distance between the coal working and the surface is only nine feet, which having fallen in, has formed an admirable air-shaft.

I may here mention, that I think it injudicious in the majority of instances to commence a coal work by means of a "day level" in the coal itself, especially as in the case now before us, for the following reasons:—1st, from the disintegrated state of the strata and the coal by the action of the atmosphere, it is very difficult to support the entrance; and the coal is not generally saleable until they have gone some considerable distance in. 2d [sic] the mine is very likely to be flooded from the surface water entering the open adit, or by an extraordinary rise in the river; whereas, if they had sunk a shaft some distance away from the outcrop, solid ground would have been obtained, the coal at once good and saleable, besides obviating the difficulties likely to be met with under the foregoing circumstances.

Mr. Williams is now driving for a new work on the deep side of the former, which promises to be carried out on a better and more economic system, and which, I doubt not, will be much more beneficial to himself and the consumer.

I cannot properly apply the term "system" in the mode adopted as yet, in working coal in this district, for it is neither "pillar and stall" nor "broadwall"[e], but merely dragging the coal out, without regard to any established rule; nothing but an easy yield, and high price, could have made the works pay.

Notwithstanding these drawbacks, too much credit cannot be given to Mr. Williams as the first adventurer in this important branch of business. I can only regret that his want of knowledge, and the difficulty of getting a competent overseer and workmen, should have rendered his profits so much less than they ought to have been, independent of the loss of much valuable fuel. I have appended a plan of the Moggil Coal Works [sic], which will show the rude manner in which the works have been carried on.[50]

Stutchbury's plan of the mine is shown in document 3.

e. Broadwall. It is curious that Stutchbury used this term because it is never heard in the industry; he was presumably referring to "long wall". This system originated in Shropshire in England, towards the end of the seventeenth century.

Williams's attempt to intersect the seam at a high level through a new entrance—thus providing workings less prone to flooding risk—had been mentioned in the press some three years before Stutchbury's visit.[51] At that time there was optimism that it would be successful but as nothing had been achieved by the time of Stutchbury's visit it can be assumed that Williams had lost heart in the development. It is significant that shortly after Stutchbury's visit (11 May and 1 June 1854 respectively) when the portions to which the new tunnel pointed came up for auction (Selections 13 and 16) the purchaser was a William Weekes and not John Williams (document 2).[52]

Over thirty years after Stutchbury's visit a second description of Moggill Mine was given by William Fryar who had recently been appointed Queensland's inspector of mines (plate 9). William Fryar was born in Willington,

Plate 9. William Fryar.

Northumberland, the son of a mining engineer.[53] There is little known about his youth and education although his unspecified scholastic attainments were worthy of special note in his obituary.[54] In 1853, at the age of twenty-five, he came to Queensland where documentation of his career is lost until 1864 when he began working part-time for the Lands Department as a licensed surveyor. Subsequently he became a partner in a sugar plantation development, entered the Queensland Parliament in 1873[55] as the member of the Legislative Assembly for East Moreton and, for a short period in 1875, was secretary of lands. In 1877 he returned to surveying before his appointment as inspector of mines in 1882.[56] These dates suggest that the mining experience so essential to the appointment must have been gained predominantly in his youth including a period between 1853 and 1858 at Moggill Mine. This phase of his career will be examined later.

In 1890 the assistant government geologist, William Rands, prepared a report on the Brookfield, Pullenvale and Moggill districts.[57] The section of the report on coal mining at Moggill is as follows:

> Coal was worked in the adjoining selection, No.15, over thirty years ago, by Mr. Fryar, the present Government Inspector of Mines, to whom I am indebted for the following notes:—
>
> A detached fragment of a seam of coal from 10 to 12 acres in extent, at the south-east corner of portion 15, was worked during about ten years from 1848 to 1858 or 1859.
>
> The seam was from 5 feet 3 inches to 5 feet 6 inches high, with a considerable band of clay shale under the top 18 inches of coal, and an occasional band of inconsiderable thickness near the bottom. It was bounded on the east by the river at too shallow a depth to be worked with safety underneath, even if it had not been displaced in the channel; bounded on the north-east by the outcrop in the elevated ridge, near the river at that point; and on south-west or south-south-west by a fault or dislocation of the strata—evidently an upthrow in that direction.The line of fault was followed, but no coal found above, so that the surface had apparently been denuded of the counter part of that seam. Other seams are found in the gullies to the southward (one of which is believed to be identical with that recently opened up in the adjoining portion), but which have not hitherto been considered capable of being profitably worked.
>
> The dip of the seam varied both in amount and direction owing to its proximity to the fault, but, generally, it may be said to have been to the south-east at about 1 in 5, or say 11 to 12 degrees.
>
> The coal was of a fairly marketable quality,[58] very similar to those now worked in the neighbourhood of Ipswich and Bundanba, and all the available supply was exhausted, there being only fragments of the perished coal near the surface left on the one side, the face of the stratum in dangerous proximity to the river channel on another, and the bared dislocation on the third. A stone drift was driven into the strata which disclosed only the ordinary blue shales of the coal measures, . . .

The Final Years of Moggill Mine

Moggill Mine did not remain long in full production after Stutchbury's visit. From its opening in 1848 output rose to 400 tonnes per year in 1850 and reached 1,320 tonnes by 1851.[59] After recording a production of 1,200 tonnes in

1858[60] it vanishes from the statistics and its final years are still shrouded in uncertainty and intrigue.

In August 1854, only a few months after Stutchbury's visit, Williams sold Portions 14 and 15 Moggill to the Hon. Louis Hope and Frederick Bigge[61] and disappeared finally from the coal mining scene, concentrating his activities on prospecting nearer the mouth of the Brisbane River but without very much success.[62] The *Moreton Bay Courier* had scented that sale negotiations were in progress, reporting in June 1854 that Williams had sold his mine for £3,500 ($7,000) to a Mr Bickerton, who had immediately re-sold it for an extra £500 profit.[63] It was inferred that the purchaser was a new arrival in the district, but it is hard to understand why even a stranger would part with such a large sum for such a doubtful asset, and the £3,500 which Williams received from the transaction made in August 1854 also included payment for two allotments which Williams had recently acquired in Yeerongpilly.[64]

The purchasers, Louis Hope and Frederick Bigge, were partners in the trading company of C.H. Marshall and Company. In collaboration with two local merchants, John Panton and Henry Buckley, plus the manager of the Ipswich branch of the Bank of Australasia, George Faircloth, they resolved to keep the mine operating but as none of them had any practical coal-mining experience they looked for someone to take charge. Their choice fell on William Fryar who had recently arrived in Queensland. Both William Rands (writing in 1890) and L.C. Ball (in 1913) mention that he worked at the "Old Moggill pits" for several years,[65] while the 1899 geological map by Cameron names the coal deposit at Old Moggill as "Fryar's Seam".[66]

Advertisements in 1856 for screened coals from "Moggill Coal Mines", and in 1857 for miners to work at the mine, both refer to William Fryar as the manager.[67] The discrepancy between McClurg's view that Williams had disposed of his coal interests to residents in the Moggill district[68] and the contemporary account that the buyer of the mine was a new arrival in the district[69] is harmonized to some extent by the hypothesis that Fryar settled in the Moggill district on his arrival from Britain and rented the mine from its new owners for a number of years or ran it on their behalf. It is known that he purchased a portion of land near the mine in 1856.[70]

The records show that the worst floods since 1842 inundated the Brisbane Valley in May 1857.[71] They caused great damage to Moggill Mine, filling the workings with water, partially destroying the wharf and washing away two or three of the men's houses.[72] The wharf was rebuilt but was again damaged by floods in the following year.

In 1890 Fryar is quoted as giving the closing date of the mine as 1858 or 1859 following inundation by a flood[73] but in a report which he submitted in 1899[74] he revised this estimate and said that it was closed after flooding in 1862 or 1863. Pugh, writing in 1864, mentioned that coal had not been worked in the vicinity of Moggill for some years,[75] while in 1876 the newly-appointed geological surveyor to Queensland, Mr A.C. Gregory, suggested by implication that it continued to operate until about 1866.[76] We know that in 1859 the partners raised a loan to prop up the operation but that a visitor found the seam nearly exhausted with only one man employed underground. In the same year a new shaft was sunk nearby but with unknown success.[77] The truth of the matter is probably that Fryar moved elsewhere in about 1860, realizing that the mine was finished and leaving behind operations which were predominantly of an exploratory or clearing-up nature. The crisis came in 1861 when it was discovered that George Faircloth had sufficiently confused his diverse roles of bank manager, coal merchant and entrepreneur to misdirect bank funds into the account of C.H. Marshall and Company. The company was wound up and its property put up for sale.[78] With the purchase of the site in 1864 for building development the story of Moggill Mine closes.

A major obstacle to continuing the mining operation had been the lack of knowledge of the detailed geology of the area and the absence of professionally-trained geologists who were capable of understanding the local complexities. Samuel Stutchbury in 1854 restricted his attention to the obvious outcrops in the district which he mapped with great care.[79] But they did not approach within a kilometre of Moggill Mine and a comparison of the geological maps published in 1899 and 1955 shows that forty years after the mine had closed a definitive picture had still not emerged (see Figure 12 on page 168). Attempts were made for many years afterwards to locate "Fryar's Seam" in adjacent blocks[80] because of its quality, which was much higher than that of the average Ipswich coal,[81] but they were of little avail and met with only meagre success.

The new owners of Portions 14 and 15 Moggill were John Panton and Henry Buckley[82] who christened the area Moggill Town and commissioned a Brisbane surveyor of doubtful reputation[83] Louis le Gould to subdivide it into over a hundred building blocks. The plan never became fully effective although some of the roads are still in existence under new names. All obvious signs of the mine itself have long since disappeared from Portion 15 Moggill but in recent years the tiny subdivisions have been recombined into substantial portions and the area has developed as one of Brisbane's most exclusive residential areas.

Difficulties in locating the precise position and extent of the workings at Moggill are that Stutchbury's plan (docu-

Plate 10. Site of Williams's Moggill Mine, with Stutchbury's plan of the workings superimposed on it. (Photographed in May 1979.)

ment 3)[84] does not relate the position of the mine to the boundaries of the portions, and the scale shown on the plan appears to be incorrect. However, a review of the available evidence has permitted a reasonable estimate to be made of the limits of the workings and these are superimposed on an aerial photograph of the site in plate 10. At one point the 1974 floods have scoured the river bank and exposed the coal seam where it dips under the water (plate 11). Currently the presence of the coal causes problems for barges in the vicinity when dredging the river bed for gravel.

The site (or what is believed to be the site) of the so-called Pioneer Mine at Redbank has been relatively less affected by the action of flood and time than Moggill Mine. In the early 1920s it was reopened by Mr Charles Macpherson who was able to crawl in for a short distance and in 1954 Denmead mentioned that the entrance could be located without difficulty although the tunnel was blocked.[85] To-day it is barely visible and will soon disappear completely through the action of natural erosion (plate 7).

Plate 11. Eroded area on the riverbank at the site of Moggill Mine. (Photographed in June 1980.)

Mining Safety

In Britain the rapid rise in coal production during the first half of the nineteenth century, which was necessitated by the vast expansion of her industry, led to an alarming increase in underground casualties. Between 1800 and 1840 over a thousand deaths were reported in major pit disasters together with an unspectacular but relentless toll exacted by relatively minor accidents. In that period no industry was more investigated, reported upon or made subject to legislation and inspection than coal mining. The first British Coal Mines Act in 1842[86] prohibited the employment of women underground but curiously no specific safety requirements were stipulated until as late as 1850, when an Act was passed which established the Mines Inspectorate and made it obligatory for the owner to maintain an accurate plan of his mine at his own expense.[87] Inspectors were appointed but there were no rules or regulations for them to enforce until 1855.[88]

Stutchbury had had an intimate involvement with these developments and it clearly influenced his approach to coal production and to safety legislation in New South Wales, which was non-existent at that time. In his twelfth report he discussed the methods of working which he considered most appropriate to the Northern Districts.[89]

The pillar and stall mode of working, the most ancient system, still continued in the majority of cases in England, is difficult to alter, from several causes, such as the proverbial dislike to innovation, and from the system having been carried to a considerable extent, rendering it very difficult and expensive to carry the alterations necessary from one system to the other into effect; but in opening new works all improvements ought to be taken advantage of; therefore it is that I would strongly recommend that the "broad wall" should be adopted, in preference to the "pillar and stall" system, for the following reasons:—

The pillar and stall ultimately involves great pecuniary loss, the pillars, upon their removal, being considerably crushed by the incumbent weight. This, combined with the more important fact of its being extremely dangerous to life, ought to be sufficient, even if it were more economical, to induce parties to adopt the safer principle.

It will only be necessary for me to point out the main difference between the two systems, to satisfy parties, not prejudiced by ancient usage, of the advantage of the one mode of working over the other.

The pillar and stall system consists in driving a "main way" through the coal in the direction of its nearest level, and cutting stalls, generally about the size that would be requisite for stabling a horse, separated by pillars of coal of sufficient size to support the roof, according to the character of the coal and roof, and thus by a series of ways and stalls working the coal to the extent of their boundary. Having attained this point, it is usual to rcmovc thc pillars, by working back to thc pit or adit, as the case may be, leaving the roof to fall behind them; but it too often happens, from the desire to mix the small coal of the pillars with the stall coal, that they remove some of the pillars before they have reached their boundaries. The consequence is, that portions of the roof fall, leaving cavities which become recipients of the carburetted hydrogen gas evolved from the workings, which, from its lightness, compared with atmospheric air, ascends into the hollows, cracks, &c. which may have been caused by the falling in of the roof, thus forming a natural gasometer, the danger arising from which is, that upon any sudden depression of the atmosphere the carburetted hydrogen by its expansion will underflow into the workings, and, mixing in a certain proportion with the common atmosphere, becomes explosive; and to this may be attributed all the dreadful explosions and loss of life so frequent in the northern and Welch [sic] districts, where this system is so universally adopted; while, on the contrary, in the Somersetshire and Gloucestershire collieries, where the "broad wall" system is carried out, there is scarcely a recorded instance of loss of life by firedamp. It is not because

> the south-western coal yields less gas than the northern; for the contrary is well known, as well by experiment as by its being largely used in the manufacture of illuminating gas.
>
> The "broad wall" method of working consists in taking a face of coal, say from ten to thirty or more yards, fetching the whole of it without leaving pillars, piling the spoil here and there in what is termed pinnacles, and supporting the tram and air ways by means of timber, of which, although an expensive item in England, there is an ample supply in this country. By this means a perfect system of ventilation may be carried on (so difficult to preserve in the pillar and stall system), preventing danger from the accumulation of gas, and therefore avoiding the necessity of employing the Davy and other safety lamps.
>
> I am induced to offer these observations from observing that the small commencement that has been made in working coal in this district is upon the northern principle; I presume in consequence of the more modern and improved mode of working not being known.

These observations were made before he had visited the Moggill Mine. After the visit he was clearly disturbed by what he had seen and in his thirteenth report[90] made two important recommendations.

> . . . the Government should reserve the right (now and for ever) of inspection by their agents, so that the works shall not only be carried out upon well regulated systems, or principles, for the protection of life, as also for the economic value of the coal to the common weal; for once lost or wasted it can never be recovered; and further, that all parties working mines should be obliged, under certain penalties, such as forfeiture of lease, or otherwise, to keep accurate plans of the workings as they are proceeded with, and forward periodically copies of the same, to be deposited as records in any place of safety that His Excellency may think proper.

The first recommendation, although admirable in principle and intention was not implemented until after the passage of the Mines Regulation Act through the Queensland Parliament in 1881. The second recommendation had an interesting sequel. The Collieries Bill was introduced into the New South Wales Parliament shortly after Stutchbury returned to Sydney and on 30 November 1854 the Coal Mines Act received Royal Assent.[91] It required that all pits, levels, shafts, tunnels and borings made for the extraction of coal should be recorded, that plans and descriptions of the operations were to be kept, and that all records had to be immediately available for inspection by a government examiner of coal mines (document 4). The terms of the Act fell far short of the British safety legislation which had been passed four years earlier but the Act gave Australia its first coal mining legislation and the first examiner, William Keane, was appointed in December 1854.[92] There is no evidence to show that the legislation was ever enforced in the North-

2878 Nos. 32 & 33. 18° VIC. 1854.

Public Parks.

No. XXXII.

COAL MINES.

An Act for the Registration and Inspection of Coal Mines in the Colony of New South Wales. [*30th November*, 1854.]

Preamble.

WHEREAS the Coal deposits of New South Wales are now being actively sought after and many of the known seams are extensively worked to satisfy the increasing demand for Coal it is expedient for the interests of the public and to prevent accidents that a record should be kept of all pits levels shafts tunnels and borings made for the extraction of Coal Be it therefore enacted by His Excellency the Governor of New South Wales by and with the advice and consent of the Legislative Council thereof as follows :—

Governor to appoint Examiner of Coal Mines.

1. The Governor of the said Colony may appoint any fit person to be an Examiner of Coal Mines and from time to time to remove such Examiner and appoint another in his stead.

Plans of the workings of Coal Mines to be kept and certified copies thereof to be furnished to the said Examiner.

2. All persons working Coal Mines shall keep or cause to be kept plans of their workings and furnish certified copies of the same to the said Examiner of Coal Mines or shall permit such Examiner either by himself his assistants or such persons as he may duly authorize to enter upon and examine all such workings for the purpose of duly recording the same by plans sections and descriptions as he may deem needful.

Owners of Collieries may demand copies of said plans on payment of a reasonable fee.

3. The owners of Collieries may demand of the said Examiner copies of all such plans sections and descriptions of their works as may be recorded in his office upon payment of a reasonable fee for the same.

Owners of Collieries refusing to the said Examiner the necessary means of examination and inquiry under this Act subject to a penalty.

4. Every owner or agent of any Coal Mine or Colliery who shall refuse to the Examiner the means necessary for making at all reasonable times any entry inspection examination or inquiry under this Act and every person who shall wilfully obstruct any such Examiner in the execution of this Act shall for every such offence be liable to a penalty of not less than ten nor exceeding twenty pounds recoverable in a summary manner before any two Justices of the Peace.

Interpretation clause.

5. The term "owner" of a Colliery or Coal Mine shall mean the immediate proprietor or lessee thereof and the term "agent" of a mine shall mean any person having on behalf of the owner of any mine the care and direction thereof.

Document 4. Australia's first coal-mining safety legislation, 1854 (chap. 3, n. 91).

ern Districts or that the area was ever visited by Keane during the period before Separation.

On the question of introducing a royalty system Stutchbury advised the government to retain the power "... to sell, lease, or let to parties who have the skill, capital, and energy, tracts of land which, properly worked, would supply for ages all that could possibly be demanded of them ..."

This wise counsel was followed by the publication of government plans for the leasing of known coal-bearing lands in the Newcastle area on a rental plus royalty basis,[93] but they did not extend as far as the Moreton Bay area and were never implemented. Procedures in Queensland for the leasing, purchase and exploitation of land believed to hold coal-mining potential were introduced on an informal basis in 1860 but it was not until 1865 that they were legally covered by any regulation.[94] These developments will be considered in a later chapter.

Notes and References for Chapter 3

1. A. Meston, *Geographic History of Queensland* (Brisbane: Government Printer, 1895), p.159.
2. *Pugh's Almanac* (Brisbane: T.P. Pugh, 1859), p.74.
3. Ibid.
4. J.J. Knight, *In the Early Days* (Brisbane: Sapsford & Co., 1895), p.64.
5. C.C. Petrie, *Tom Petrie's Reminiscences of Early Queensland* (Brisbane: Watson & Ferguson, 1904), p.257.
6. New South Wales Archives, Location 4/2460.3, Ref. 37/10899, filed with 39/9639, para.6.
7. Ibid., Location 4/3795, letter No. 38/12, colonial secretary's office.
8. N.S. Pixley, "The History of the A.U.S.N. Co. Ltd. and its Predecessors", *Journal Historical Society of Queensland* 5 (1954): 969-1003.
9. King Papers, Vol.2 (Mitchell Library Ref. A.1977), p.105.
10. *Historical Records of Australia*, Vol.22 (Canberra: Library Committee of Commonwealth Parliament, 1924), p.731.
11. King Papers, Vol.2, p.105.
12. W. Johnston, "Old Bulimba" (note by W. Clark), *Journal Historical Society of Queensland* 1 (1918): 314-18.
13. *Votes and Proceedings of the Legislative Council of New South Wales*, 1854, Vol.2, pp.1383-95.
14. *Pugh's Almanac*, 1859, p.74.
15. J.H.C. McClurg, *Historical Sketches of Brisbane* (Brisbane: Library Board of Queensland and Royal Historical Society of Queensland, 1975), p.10.
16. A.G. Davies, "Genesis of the Port of Brisbane", *Journal Historical Society of Queensland* 2 (1935): 274-87.
17. Johnston, "Old Bulimba" (note by Clark), pp.314-18.
18. *Australian Dictionary of Biography, 1788-1850*, Vol.2 (Carlton: Melbourne University Press, 1967), pp.600-601.
19. Johnston, "Old Bulimba" (note by Clark), pp.314-18.
20. McClurg, *Historical Sketches of Brisbane*, p.10.
21. *Moreton Bay Courier*, 26 July 1851.
22. *Pugh's Almanac*, 1859, p.74.
23. Petrie, *Reminiscences of Early Queensland*, p.257.

24. *Votes and Proceedings of the Legislative Council of New South Wales*, 1854, Vol.2, pp.1383-95.
25. W. Fryar, *Annual Report to Department of Mines, 1899* (Brisbane: Government Printer), pp.156-64.
26. A.J. Norman, *Black Diamonds–the Story of Coal in Queensland* (Brisbane: Queensland Coal Board, 1975).
27. Queensland Government Titles Office (Old System), Conveyance Book 7, No.344. Also Book of Extracts, Vol.1, No.16, Conveyance 18 December 1843.
28. Knight, *In the Early Days*, p.64; *Moreton Bay Courier*, 26 July 1851; *Australian Dictionary of Biography, 1788-1850*, Vol.2, pp.600-601; and Fryar, W., "The Development and Progress of Mining and Geology in Queensland", *Proceedings Australian Association for the Advancement of Science*, Vol.4 (1895), p.363.
29. Queensland Government Titles Office (Old System), Conveyance Book 7, No.344.
30. *Votes and Proceedings of the Legislative Council of New South Wales, 1855*, Vol.2, pp.1117-22.
31. I. Jones, "The Floods on the Brisbane River", *Journal Historical Society of Queensland* 2 (1935): 288.
32. *Moreton Bay Courier*, 26 July 1851.
33. Ibid., 13 March 1860.
34. Ibid., 8 August 1846.
35. *Historical Records of Australia*, Vol.23, p.468.
36. *Votes and Proceedings of the Legislative Council of New South Wales*, 1847, Vol.2, pp.335-92; and Vol.1, p.491.
37. *Historical Records of Australia*, Vol.26, p.732.
38. *British Parliamentary Papers (Australia), 1840-1841* (Shannon: Irish University Press, 1970), pp.171-74.
39. *New South Wales Government Gazette*, 1850, Vol.1, p.143.
40. *Moreton Bay Courier*, 26 February 1848.
41. Fryar, Report to Department of Mines, 1899, pp.156-64; and *Moreton Bay Courier*, 11 November 1848.
42. Queensland Government Department of Mapping and Surveying, Brisbane, Map M1118.2; and New South Wales Archives, Location 4/5432, surveyor-general's office, 1846-49.
43. Queensland Government Titles Office, Vol.30, Folio 209-10.
44. Acts 5° and 6° VIC. Chapter 36, 22 June 1842.
45. Johnston, "Old Bulimba" (note by Clark), pp.314-18.
46. *Australian Dictionary of Biography, 1851-1890* Vol.6 (Carlton: Melbourne University Press, 1976), p.216.
47. *Moreton Bay Courier*, 15 September 1851.
48. *Votes and Proceedings of the Legislative Council of New South Wales*, 1854, Vol.2, pp.1383-95.
49. *Moreton Bay Courier*, 17 December 1853.
50. *Votes and Proceedings of the Legislative Council of New South Wales*, 1854, Vol.2, pp.1383-95.
51. *Moreton Bay Courier*, 26 July 1851.
52. Queensland Government Titles Office (Old System), Conveyance Book 3, No.205, 11 May 1854 and Conveyance Book 4, No.3, 1 June 1854.
53. *Australian Dictionary of Biography, 1851-1890* Vol.4, p.223.
54. *Queensland Government Mining Journal* 5 (1904): 309.
55. *Queensland Government Gazette*, 1873, Vol.14, p.1935.
56. The *Australian Dictionary of Biography* claims that Fryar was Queensland's first inspector of mines but this title undoubtedly belongs to George Lukin who held the title briefly in 1874 (see *Queensland Government Blue Book for 1874*).
57. W.H. Rands, "The Brookfield, Pullenvale and Moggill Districts", *Queensland Geological Survey*, Report 63 (Brisbane: Government Printer, 1890).
58. A test on a sample was made in 1855 with the following results: ash 7 per cent, volatiles 32 per cent, sulphur 0.508 per cent. It was de-

fined as highly bituminous with a coke of excellent quality. See *Moreton Bay Courier*, 10 November 1855.
59. Sir Charles Fitzroy to Sir I.S. Packington, Despatch 124, 4 August 1852 (Mitchell Library Microfilm 637).
60. *Statistics of New South Wales*, 1859, p.84.
61. Queensland Government Titles Office (Old System), Book of Extracts, Vol.3, No.20.
62. Annual Progress Report for 1896-98, *Queensland Geological Survey*, Report 143 (Brisbane: Government Printer, 1899), pp.6-11.
63. *Moreton Bay Courier*, 3 June 1854; and 6 May 1854.
64. Queensland Government Titles Office (Old System), Book of Extracts, Vol.3, No.20.
65. ANZ Archives, Ipswich Branch Records, 1861, Rands, "The Brookfield, Pullenvale and Moggill Districts"; and L.C. Ball, "Prior's Pit, Moggill", *Queensland Government Mining Journal* 14 (1913): 301-2.
66. W.E. Cameron, "Geology of the West Moreton or Ipswich Coalfield", *Queensland Geological Survey*, Publication 147, Map No.2 (Brisbane: Government Printer, 1899).
67. *North Australian*, 9 December 1856; and 27 May 1857.
68. McClurg, *Historical Sketches of Brisbane*, p.10.
69. *Moreton Bay Courier*, 3 June 1854.
70. Queensland Government Titles Office (Old System), Conveyance Book 6, No.120, 7 August 1856.
71. Jones, "The Floods on the Brisbane River", p.288.
72. *Moreton Bay Courier*, 30 May 1857.
73. Ibid., 6 October 1858.
74. Fryar, Annual Report to Department of Mines, 1899, pp.156-64.
75. *Pugh's Almanac*, 1865, p.195.
76. *Votes and Proceedings of the Legislative Assembly of Queensland*, 1876, Vol.3, pp.387-94.
77. ANZ Ipswich Records, 1859; and *Moreton Bay Courier*, 5 February 1859.
78. ANZ Ipswich Records, 1862.
79. *Votes and Proceedings of the Legislative Council of New South Wales*, 1854, Vol.2, pp.1383-95.
80. Rands, "The Brookfield, Pullenvale and Moggill Districts".
81. *Queensland Coals*, 5th ed. (Brisbane: Queensland Coal Board, 1975).
82. Queensland Government Titles Office, Vol.30, Folio 209-10.
83. *Brisbane Courier*, 26 November 1863.
84. *Votes and Proceedings of the Legislative Council of New South Wales*, 1854, Vol.2, pp.1383-95.
85. A.K. Denmead, "The West Moreton (Ipswich) Coalfield", *Queensland Geological Survey*, Publication 279 (Brisbane: Government Printer, 1955), p.101.
86. New South Wales Acts 5° and 6° VIC., Chapter 99, 10 August 1842.
87. New South Wales Acts 13° and 14° VIC., Chapter 100, 14 August 1850.
88. R. Nelson Boyd, *Coal Mines Inspection* (London: W.H. Allen & Co., 1879).
89. *Votes and Proceedings of the Legislative Council of New South Wales*, 1854, Vol.2, pp.1375-81.
90. Ibid., Vol.2, pp.1383-95.
91. New South Wales Act 18° VIC., No.32, 30 November 1854.
92. *New South Wales Government Gazette*, Vol.2, 28 December 1854, p.2741.
93. Ibid., Vol.1, 21 February 1854, p.389; and Vol.1, 4 May 1855, p.1293.
94. *Votes and Proceedings of the Legislative Assembly of Queensland*, 1865, pp.1207-8.

Coal Mining in the Ipswich Area before Separation

4

Although outcrops of coal were mapped above and below "Limestone Station" on the Bremer River by Cunningham in 1828 it is not known exactly when mining started in the area. If wood had been in short supply the coal could have been used for limestone burning at any time after Logan established the kilns in 1827[1] but the claim that coal was mined in the Tivoli area for the penal settlement at Moreton Bay has already been mentioned. Contemporary records which establish the extent of its use are very meagre, and it is necessary to move forward a whole decade to Major Cotton's letter to Sydney in 1837[2] in order to find the first clear evidence that coal mining was being seriously contemplated in the area. Progress was slow for the reasons already considered but matters moved another step forward in 1843 when the Moreton Bay correspondent of the *Sydney Morning Herald* reported that:

> I have every reason to believe that we shall shortly be able to supply the steamer with coals, the produce of the district, a very fair seam of that valuable mineral having been discovered in the creek at Limestone; in fact, Mr. Moore, the blacksmith and wheelwright residing there, assured me that he works no other coal but that in his forge. A few tons as a specimen will shortly be sent to Brisbane, when a fair trial can be made of its burning qualities.[3]

The result of the test is unknown but we now know that the first coal produced in the district for regular sale came from Redbank and not from Ipswich. In order to establish definite evidence of mining on the Bremer River it is necessary to move forward another five years to 1848 when Map M1118.18[4] locates a "coal works" on common land on the west side of Woodend Pocket at the highest point which could be reached by punt on the river (document 5).

One of the actions which followed the decision by the government in 1848[5] to give the utmost encouragement to the mining of coal by removing licensing requirements for operations on alienated land was the subdivision of coal-bearing areas of crown land near to Ipswich into minute portions of a little over one hectare each, the first of these being in what is now known as Woodend Pocket (document 5). This action came in for severe criticism some years later[6] when it was claimed to have been a major deterrent to the systematic exploitation of the various coal-bearing areas around Ipswich because the individual portions had been too small to support viable operations.

The Woodend Pocket portions were released in 1848 as "coal allotments" at an upset price of £3 per acre ($15 per hectare), and were quickly sold off. But the purchases were not followed by the large increase in coal mining in the Pocket for which the government had hoped. Six years later in early 1854 when Samuel Stutchbury was making his inspection of the coalfields of the Moreton Bay area and was following down the south bank of the Bremer River he found only one active operation upstream from Ipswich: ". . . the first coal visible is in a little creek belonging to or adjoining Mr. M'Intyre's purchased land . . . The upper nine feet of coal is now being inefficiently worked by one white man, aided by a black, who brings it out in a wheel-barrow".[7] This corresponds to the position of the "coal works" on Map M1118.18 in Woodend Pocket, to which reference has already been made (document 5). Stutchbury also noticed that several openings had been made near the "coal falls" which at that time were located at the top of the Pocket immediately upstream of the confluence with the present Mihi Creek, but they were not being worked. In contrast to the inactivity on the privately-owned allotments at Woodend, coal continued to be worked on crown lands near Ipswich without the issue of a licence or any other authority. This practice was rife in the parish of Chuwar adjacent to the mouth of the present Tivoli Creek which was convenient both for loading onto the river boats and for selling to one of the boiling-down establishments. Commissioner Barney on one of his regular tours outside Brisbane reported to the colonial secretary's office in late 1850, and again in early 1851, that he had discovered three such unauthorized mines on the Bremer River and requested advice on what should be done about them.[8] Although the positions of two of them are speculative, the third can be identified from Barney's description as the Reid and Boyland venture which will now be described.

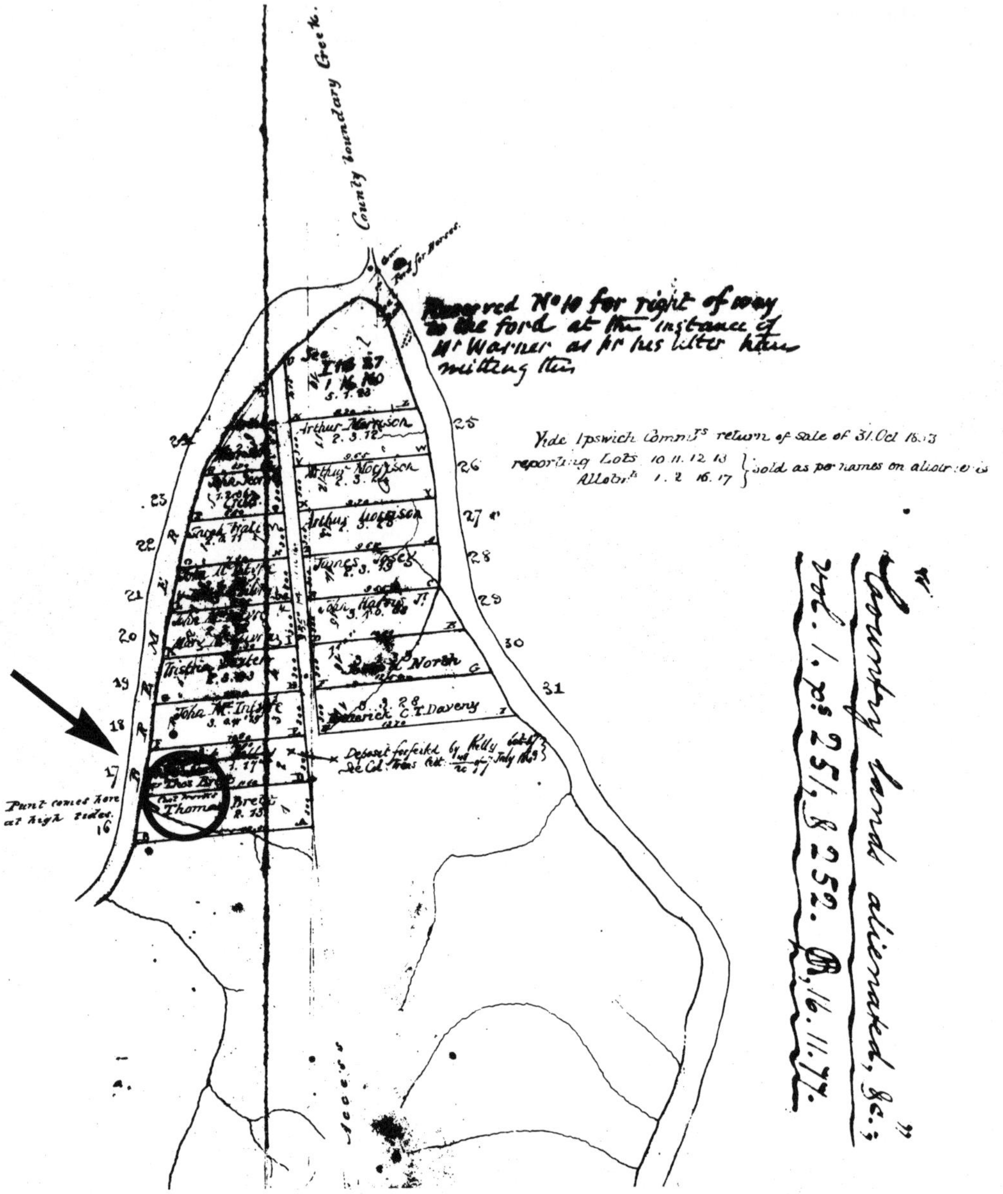

Document 5. Part of the 1848 subdivision plan of Woodend Pocket (chap. 4, n. 4), showing the "coal works" on Portion 16 which Stutchbury saw operating. See also fig. 4, p. 52. (Plan supplied by the Surveyor-General, Queensland, and reproduced by arrangement with the Queensland Government. CROWN COPYRIGHT RESERVED.)

Reid and Boyland's Mine

After 1845 the licence for operating the Russell Street ferry in Brisbane was advertised and auctioned annually. In 1847 it was bought by John Boyland and James Reid[9] who commenced a punt service to Ipswich in the same year and built a depôt there in order to take advantage of the growing wool trade.[10] When the *Experiment* sank at its moorings in Brisbane on 20 January 1848[11] it was the partnership of Reid and Boyland which immediately bought her, raised her and put her back on the Ipswich service until 1849 when she was dismantled.[12] In the same year a new boat, the *Hawk*, was built for them in South Brisbane[13] using the engine from the *Experiment*, and she continued the service to Ipswich through the 1850s and 1860s, first in association with a second locally-built steamer, the *Swallow*,[14] and later with a growing number of other steamboats (plate 6). These boats required a small but regular supply of fuel and stopped at first at Moggill Mine to load it.[15] But demand steadily increased and by 1851 the two boiling-down works in Ipswich were calling for four hundred tonnes of coal a year which, when added to the river steamer coal, was sufficient to provide a regular market for an Ipswich supplier.[16]

In 1851 Reid and Boyland supplemented their other business enterprises by opening a mine on Crown Land on the west bank of the Bremer River about two and a half kilometres downstream from Ipswich. It was later recorded that they employed "a miner holding a licence for that purpose"[17] who had no doubt been licensed and engaged in order to legalize their operation following a visit from Commissioner Barney. A reporter who inspected the site in the same year wrote that:

> This mine is about 150 yards from the bank of the creek, alongside which the steamboats can lie to take in coal or other cargo. The mine is only opened for a few feet from the side of a neighbouring hill, and a thick seam of fine bright coal has been worked into, while another is already for use underfoot. In fact the coal crops out through a large extent of the neighbouring gully as well as in the side of the ridge, and all that is required is the removal of a few feet of earth to expose the seam in almost any spot around.[18]

The description locates the mine on Crown Land immediately to the west of the present Tivoli Creek (figure 3).

Meanwhile the government in Westminster was calling for " . . . a special report on the extent and nature of the coalfield existing in the locality with a view to the subdivision of the land into allotments for sale in such a manner as would be likely to prevent a monopoly in this important article".[19] Undeterred by the failure of the policy adopted in Woodend Pocket to encourage coal mining, the colonial government resolved to repeat the experiment downstream from Ipswich. Subdivision plans were drawn up for the north bank of the Bremer River below the present Tivoli Creek where the Garden seam outcropped near the river,[20] although the coal-bearing area lying across the creek and including Reid and Boyland's mine was excluded, possibly because of the poor quality of the coal (document 6). In 1852 the upset price at auction was fixed at £7 per acre ($35 per hectare)[21] which was a high rate compared with the usual £1 per acre for agricultural land.

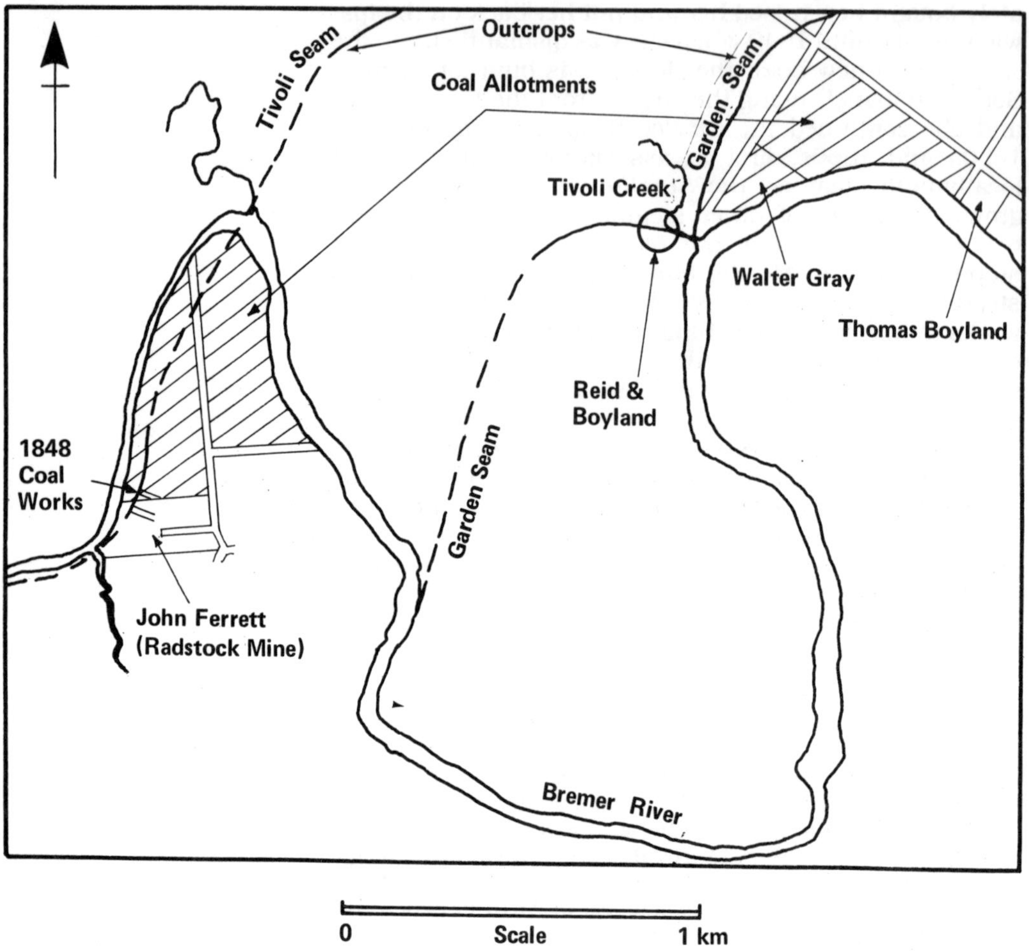

Figure 3. Early coal mines in the Ipswich area. Locations of the Tivoli and Garden seams are taken from Cameron's 1899 maps (chap. 4, n. 49). Seams not mentioned in the text are omitted for clarity. See Frontispiece for location of the map.

Messrs Reid and Boyland made successful bids for allotments 10 and 11 but later in the same year they dissolved their partnership, closed the mine and sold the *Hawk* to Captain Thomas Boyland.[22]

Samuel Stutchbury described the old mine briefly during his visit in 1854. Before setting out from Sydney he had received a tracing of the coal allotments at Tivoli which was almost certainly taken off Map I1207.20 (document 6).[23] It marks coal as occurring adjacent to the wharf reserve at the mouth of the present Tivoli Creek although it does not mention that Reid and Boyland had been working it.

> Observing on the tracing referred to (the only one sent me), that there was a wharf reserve, upon examining which, and crossing the small creek I found the remains of an old coal work (the coal being more than eleven feet thick, divided by a few thin seams of fire clay) . . . The coal has not been bottomed, and therefore may prove to be of greater thickness . . . .

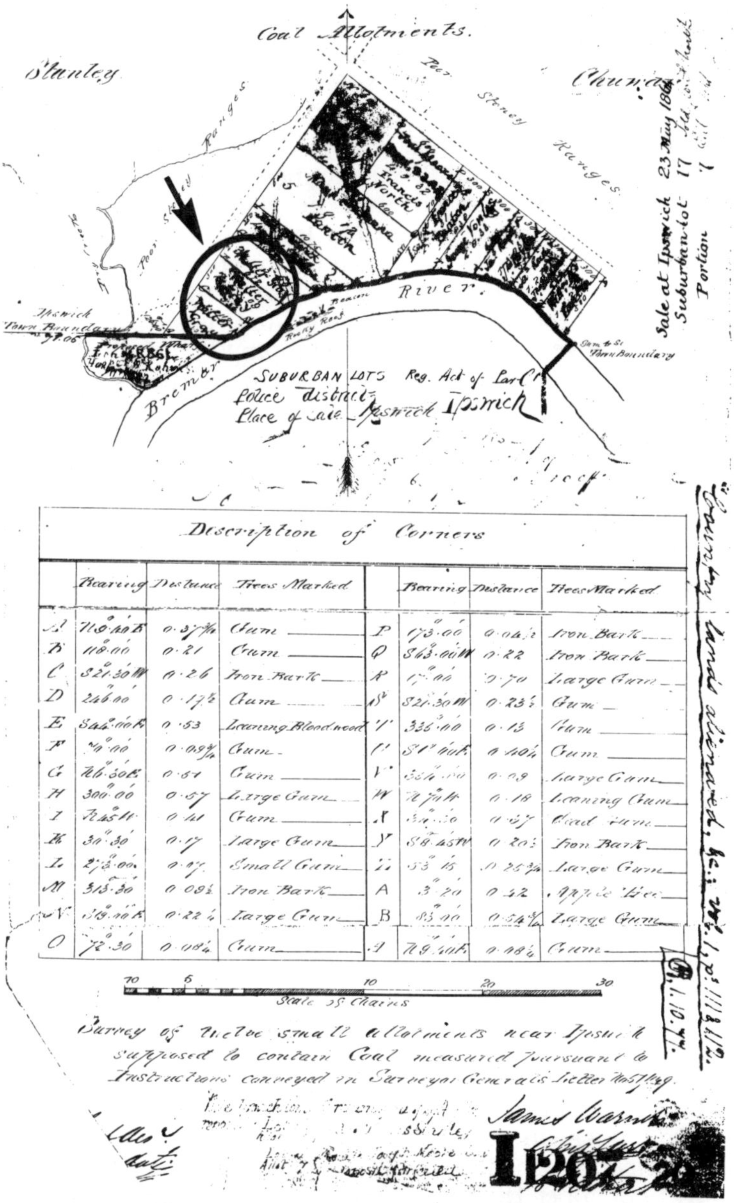

Document 6. The 1851 subdivision plan of the area immediately downstream from Tivoli Creek (chap. 4, n. 20), showing Walter Gray's allotments. (Plan supplied by the Surveyor-General, Queensland and reproduced by arrangement with the Queensland Government. CROWN COPYRIGHT RESERVED.)

From this a canal or tramway, one hundred and ninety yards in length, would deliver coal for shipment into lighters. Nearly opposite the above, in the creek, there is a seam of coal two feet thick, bearing north seventy degrees west, the creek for some extent follows the course of a line of fault. About one mile north-east of the thick coal above described, it may again be seen in several of the ramifications of the creek, while the ranges north of the "coal allotments" are strewed with nodular iron ore, resulting from the breaking up of the denuded seams.[24]

Stutchbury included in his report a sketch of the proposed canal or tramway for linking the mine with the Bremer

River but the suggestion does not seem to have been pursued possibly because the bright appearance of the coal belied its inferior quality.

The new owner of the *Hawk*, Captain Thomas Boyland, was already the owner of a piece of land immediately downstream from the "coal allotments" (figure 3). In 1853 a newspaper reported that he had "... opened an excellent seam of coal on his land on the bank of the Bremer River and is working it for the supply of his steam boat *Hawk*".[25] The Rev. Henry Stobart in recalling his journey in the *Hawk* from Ipswich to Brisbane in June of the same year refers to them stopping "... half way to take in coal from a pit mouth close to the river's edge".[26] It is unlikely that this stop was made at Moggill because elsewhere in his account he quotes the Rev. W.B. Clarke—a noted geologist of the period—as saying that northwards for a long distance from the pit was "one vast coalfield". This could only refer to the area extending through the Tivoli coal allotments.

Captain Boyland persevered with his mine and a year later, in 1854, when coal was short he collaborated with one of the Pettigrew brothers in working it.[27] In 1854 there were two brothers with the unusual name of Pettigrew in the Moreton Bay area. The elder, William Pettigrew, was a mechanic who had emigrated to Brisbane on the *Fortitude* in 1849, when twenty-four years of age. After working for some years as a surveyor for Commissioner Simpson of Woogaroo, near Redbank,[28] he bought a block of land in South Brisbane and by 1854, with his partner David Kergow, was struggling to put Queensland's first steam-driven sawmill into operation.[29] He was so deeply involved in starting up the new mill that it would be surprising if he could have spared the time to open a mine at Ipswich although this possibility cannot be completely ruled out because one of his great problems in 1854 was maintaining an adequate supply of coal.[30]

William Pettigrew's younger brother, John, joined his brother and David Kergow in the sawmill partnership in 1853, but soon withdrew because of his lack of interest in engineering which led to friction between the partners.[31] By September 1854 he had bought a small subdivision in Ipswich,[32] subsequently opening a store there and ultimately becoming one of its leading citizens. In the period between the dissolution of the partnership with his brother and David Kergow, and his opening of a store in Ipswich, John Pettigrew could have collaborated with Captain Boyland. In any case the mining operation does not appear to have survived for long and no production figures are known.

Reporting on the early days of coal mining in the district a writer summarized the elementary procedures adopted at the time: "Colliers then set to work on some of the outcrops appearing on the sides of the steepest hills. Tunnels were soon made, and the mineral was wheeled out by means of barrows. As the tunnels ran almost horizontally, no mechanical aid was required in bringing the small waggons from the underways".[33] The primitive and temporary nature of these early operations is obvious from this description.

Radstock Coal Pit

The coal famine of early 1854 referred to in chapter 2 had at least one important and beneficial effect on the Northern Districts; it stimulated the opening of additional mines situated in positions where their owners felt they could supply the market and which were free from periodic flooding by the Bremer or Brisbane rivers. The owner of one of them was Mr John Ferrett.

John Ferrett was a West Country man, born in Dorset in England in 1812 and moving to Ipswich in the early 1840s where he established the Wallam Station near Ipswich situated on the creek of the same name (plate 12).[34] In the 1850s he turned his attention to coal mining, purchasing three portions immediately to the south of the Woodend Pocket coal allotments (figure 3).[35] Writing in 1899 Fryar refers to " . . . a seam opened in the early days of the colony on the land of Mr.Ferritt [sic] within the boundary of the town of Ipswich. This cropped out on the river bank, dipped at a high angle, and formed part of the formation of a peninsula or pocket jutting into a bend of the river . . . "[36]

John Ferrett called his first enterprise the Radstock Pit, presumably after the nearest coal mining town to his birthplace. Radstock is a small English town, in the county of Somerset, where coal was discovered in the late eighteenth century. The pit was located on Portion 118 Ipswich where "Ferrelt's [sic] Tunnel" is shown on the map accompanying the 1955 Geological Survey of the West Moreton (Ipswich) Coalfield.[37] Operations began in the first half of 1854[38] and by November of the same year the *Moreton Bay Courier* was reporting "We have seen a sample of coal from the Radstock coal pits above the falls on the Bremer River, a mile or two above Ipswich. This mine has now been for some time in working order and has yielded about 400 tons of coal since it was opened seven or eight months ago".[39] It was stated that the owner was considering starting an export trade in the coal, and an agent was being sought to transport it to Brisbane. However, the hopeful start which the pit made was not maintained when the famine had passed. Forty-five years later Fryar, reflecting on the adverse influence of transportation costs on the feasibility of a mining operation, cites Ferrett's mine at Ipswich amongst his examples.

Plate 12. John Ferrett. (Reprinted, by permission, from M. J. Fox, *History of Queensland*, Vol. 1 Brisbane: State Publishing Co., p. 468.)

> . . . then, as now, there, as elsewhere, the value of coal depended on its geographical situation, and the facilities available for getting it to the place of consumption. The demand was limited; the city of Brisbane was the chief port and principal consumer, Moggill was the nearest outcrop, and the cost of conveyance down the river was as high as the price of coal at the river bank is to-day.[40]

So although the pit continued to be worked and was quoted some years later as an example of a rich coal mine[41] it was a small operation and intermittent in its output.

John Ferrett was not discouraged by the transport difficulties and later had significant interests in Woodend Coal Mines[42] which were situated in the pocket of the same name and adjacent to his first operation.

Walter Gray's Mine

One of the most successful Ipswich coal mining operations of the 1850s was Walter Gray's mine. Walter Gray was born in Elgin in Scotland in about 1812, the illegitimate son of a local artisan.[43] His connection with Ipswich dates back to 1847 when he opened a store and set up as a wool agent.[44] He was described as a shrewd but just businessman, blunt in manner but charitable in disposition.[45] Throughout his life Gray's energies were concentrated in commerce and business and he found little time to participate in local government or parliamentary matters although he took an interest in education, and was an establishing member of the Northern Australian Jockey Club in Ipswich. He became an active and energetic promoter of the direct shipment of produce from Queensland to Great Britain, involving himself in the speculative and short-lived Moreton Bay Steam Navigation Company in 1852, and two years later in the equally ill-fated Moreton Bay Immigration and Land Company. The coal famine in early 1854 led to the floating of the Moreton Bay Coal Company in Ipswich, with Walter Gray as one of the members of the provisional committee.[46] The company came to nothing but as Gray already owned three adjacent coal allotments at Tivoli Creek, which he had purchased in June 1852,[47] he decided to open his own mine. Fryar, writing in 1899 and discussing the operations of the Tantivy Mine which was active at that time, goes on to say that "The coal on this estate has been worked periodically almost from the beginning of the industry of coal-mining in Queensland; it adjoins the mine formerly worked by Mr. Gulland, which is again adjoining that opened by Mr. Walter Gray in 1854".[48] Following these clues, a map contemporary with Fryar's account[49] clearly places the 1854 mine on Walter Gray's allotments (document 6).

A curious omission from Stutchbury's reports was any reference to Walter Gray's mine. His visit took place in the first half of 1854 when shaft sinking must almost certainly have been proceeding, if indeed it had not been completed. As the first vertical shaft in the area it should have attracted special mention. One possibility is that the opening date given by Fryar and generally accepted as correct[50] is untrue. We have already seen that Fryar's date for the closure of Moggill Mine shows some discrepancies, and a contemporary account suggests that Walter Gray's mine might not have opened until 1855.[51]

The shaft intersected what later was called the Garden Seam[52] and which had been worked some years earlier by Reid and Boyland on the western side of Tivoli Creek. The Garden Seam to the east of Tivoli Creek ". . . furnished an excellent steam coal . . . " and was reported in 1899 as having an ash content of about 13 per cent.[53] This was inferior to the best Newcastle coals which assayed only 7 per cent ash[54] but was better than the usual run of Ipswich coals.[55]

For a contemporary account of the location, layout and

operation of the mine we are indebted to a reporter of the *Moreton Bay Courier*. Even after allowance is made for journalistic enthusiasm the picture presented is in striking contrast to the wretched conditions which Stutchbury had found at Moggill a couple of years earlier.

> The mine is situated on the Bremer, about 50 yards from the river bank. The shaft is 100 ft. in depth, 50 ft. being through a layer of hard sandstone which lies immediately over the coal and affords a secure roof to the underground workings. The layer of sandstone dips towards the river under which it passes; and the workings thus protected by a wall of solid stone, are preserved in a comparatively dry state, notwithstanding their proximity to the river. The seam of coal is almost six feet in thickness and of superior quality. The underground workings extend from both sides of the shaft to a considerable distance and from these main passages also diverge branch workings, which meet and thus receive proper ventilation. The coal is conveyed to the shaft from the place where it is excavated in boxes which run on tramways of a very simple and inexpensive construction. The boxes are raised to the mouth of the pit by means of a machine constructed by Mr. Petrie, of Brisbane,[56] and are then conveyed by another tramway to the wharf where they can be tipped, either at once into the boats lying alongside or onto the wharf as a stock for future consumption. The buildings above ground consist of a large store, overseer's house, extensive dwelling house for the workmen, blacksmith and carpenter's shops, etc. At present there are but few men employed and the quantity of coal raised is about 100 tons per week; but no doubt this quantity will be much increased when the arrangements now in progress are completed, by means of which the proprietors hope to be able to supply Brisbane at prices very little above those now charged at the pit mouth.
>
> We have been told that its quality is fully equal to Newcastle coal.[57]

Walter Gray's mine prospered in the second half of the 1850s. It withstood the disastrous floods of May 1857 which ruined the Moggill Mine but it was put out of action by the October flood in 1858 when the whole of the workings were filled with water and the tramway destroyed.[58]

In 1857 Walter Gray appointed William Pettigrew as Brisbane agent for coal from his pit[59] and in the following year it was the major source of coal in the Northern Districts, producing some 2,200 tonnes[60] despite the flooding already mentioned. In 1860 the output approached 5,000 tonnes[61] but it had been overtaken by Redbank Mine and production declined rapidly, ultimately ceasing two years later,[62] reputedly as a result of the workings being flooded again[63] but more probably because of the limited coal reserves or the death of Walter Gray which occurred in April 1862.[64] The unfortunate fact of Gray's illegitimacy only became known after his death and it led to a successful disputation of his will, resulting in the forfeiture of his wife's rights to the estate and the reversion of control of the mine site to the mortgager, William Boyd, to whom

Gray had sold the freehold in 1859.[65] No evidence has been found to suggest that the mine was ever reopened.

Coal mining procedures in the Ipswich area before Separation were summarized by Stutchbury after his visit in 1854. He commented that " . . . many scratchings (I cannot call them workings) have been made at various times, in the bank for coal, and deserted, but why, I cannot imagine, as there is an abundance of good coal which merely requires to be brought out".[66] In making this assessment Stutchbury failed to recognize that many Ipswich coals were extremely friable and possessed an unacceptably high ash content so that seams which at first glance appeared to be well worth mining were in fact unsuitable for commercial use. This important fact became increasingly obvious as the years went by and led to an hiatus in coal production in Ipswich for a few years after 1862 until Hooper and Robinson commenced their operations in 1866.

To-day, Woodend Pocket is littered with the debris of later coal operations but on the river bank traces can still be seen of the mine worked, according to Stutchbury, by " . . . one white man, aided by a black . . . ".[67] The sandstone capping rock over the portal has fractured but the terrace across which the coal was hauled to the waiting barges or lighters on the Bremer River is clearly visible (plate 13). The entrance to Ferrett's Radstock Mine has col-

Plate 13. The mine on Woodend Pocket visited by Stutchbury in 1854. The entrance is left centre. (Photographed in June 1980.)

Plate 14. Radstock Mine on Woodend Pocket. The entrance lies immediately below the tall tree. (Photographed in June 1980.)

lapsed long-since but its position can be easily located by a watercourse which leads from it towards the river (plate 14). No information on the extent of the underground workings of either mine has come to light. The plans of the New Ebbw Vale colliery which worked the Woodend Pocket extensively in later years are completely blank in the area of the two old mines.

In addition to surface changes caused by exposure to periodical flooding, the area round the mouth of Tivoli Creek has been redeveloped from time to time since 1860. The latest changes include the relocation of the road bridge, the installation of a pumping station on the old wharf reserve and the closure or relocation of some streets in its immediate vicinity. However the situation of the outcrop on which Reid and Boyland worked can be tentatively identified in a gully below the back gardens of houses in Hope Street.

The three allotments which included Walter Gray's mine were purchased in 1871 by Robert Archibald,[68] the manager of Old Tivoli Pit. Later a tramway from the pit, to which reference will be made in chapter 9, ran across them on its way to a wharf situated on the bank of the Bremer River immediately below Tivoli Creek. Moreover, in the present century the foundations of Bank Street were laid across the allotments. The first successful coal mine in Ipswich has therefore disappeared without trace under subsequent development.

Notes and References for Chapter 4

1. J.G. Steele, *Brisbane Town in Convict Days, 1824-1842* (St. Lucia: University of Queensland Press, 1975), p.73.
2. New South Wales Archives, Location 4/2460.3, Ref. 37/10899, filed with 39/9639, para. 6.
3. *Sydney Morning Herald*, 22 August 1843.
4. Queensland Government Department of Mapping and Surveying, Brisbane, Map M1118.18.
5. *Historical Records of Australia*, Vol.26 (Canberra: Library Committee of Commonwealth Parliament, 1925), p.732.
6. *Brisbane Courier*, 24 October 1861.
7. *Votes and Proceedings of the Legislative Council of New South Wales*, 1854, Vol.2, pp.1383-95.
8. Commissioner Barney to Colonial Secretary, 23 September 1850 (Mitchell Library Microfilm A.227/629); and 26 January 1851 (Mitchell Library Microfilm A.227/632).
9. L.E. Slaughter, *Ipswich Municipal Centenary* (Ipswich: City Council, 1960).
10. *Moreton Bay Courier*, 15 January 1848.
11. Ibid., 22 January 1848.
12. Ibid., 29 September 1849.
13. Ibid., 8 September 1849.
14. Slaughter, *Ipswich Municipal Centenary*.
15. *Moreton Bay Courier*, 8 August 1846.
16. Ibid., 4 October 1851.
17. Ibid., 8 August 1846.
18. Ibid., 4 October 1851.
19. Mitchell Library Microfilm A.227/632.
20. Queensland Government Department of Mapping and Surveying, Map I1207.20.
21. Ibid.
22. *Moreton Bay Courier*, 3 July 1852; and 12 February 1853.
23. Queensland Government Department of Mapping and Surveying, Map I1207.20.
24. *Votes and Proceedings of the Legislative Council of New South Wales*, 1854, Vol.2, pp.1383-95.
25. *Moreton Bay Courier*, 23 April 1853.
26. H. Stobart, "Journal of a Visit to New South Wales". Mitchell Library, MS.1229, p.99.
27. *Moreton Bay Courier*, 3 June 1854.
28. Ibid., 12 August 1850.
29. W. Pettigrew, Personal Papers. Royal Historical Society of Queensland, Newstead House, Brisbane.
30. Ibid.
31. Ibid.
32. *Moreton Bay Courier*, 16 September 1854.

33. *Ipswich and Bundamba, West Moreton, Series No.1* (Brisbane: Government Printer, 1908).
34. D.B. Waterson, *A Biographical Register of the Queensland Parliament*, 1860–1929 (Canberra: Australian National University Press, 1972), p.55.
35. Queensland Government Department of Mapping and Surveying, Map MT23.
36. W. Fryar, Annual Report to Department of Mines, 1899 (Brisbane: Government Printer), pp.156–64.
37. A.K. Denmead, "The West Moreton (Ipswich) Coalfield", *Queensland Geological Survey*, Publication 247 (Brisbane: Government Printer, 1955), Map 3.
38. *Moreton Bay Courier*, 3 June 1854.
39. Ibid., 11 November 1854.
40. Fryar, Annual Report to Department of Mines, 1899, pp.156–64.
41. *North Australian*, 13 January 1857.
42. Waterson, *Biographical Register*, p.55.
43. Queensland Government Titles Office (Old System), Special Grants 2C, Folio 11, 2 August 1867, Deed of Grant 17396.
44. Slaughter, *Ipswich Municipal Centenary*.
45. *Brisbane Courier*, 10 April 1862.
46. *Moreton Bay Courier*, 30 August 1851; 10 July 1852; 21 October 1852; 6 May 1854; and 27 May 1854.
47. Queensland Government Department of Mapping and Surveying, Map I1207.20.
48. Fryar, Annual Report to Department of Mines, 1899.
49. W.E. Cameron, "Geology of the West Moreton or Ipswich Coalfield" *Queensland Geological Survey*, Publication 147 (Brisbane: Government Printer, 1899).
50. D.C. Mengel, "Ipswich Coal-Field, Queensland", *Economic Geology of Australia and Papua New Guinea*, Vol.2, ed. D.M. Traves and D. King (Melbourne: Australasian Institute of Mining and Metallurgy, 1975), p.290.
51. *Moreton Bay Courier*, 29 November 1856.
52. *Queensland Coals*, 6th ed. (Brisbane: Queensland Coal Board, 1978).
53. Cameron, "Geology of the West Moreton or Ipswich Coalfield".
54. Fryar, Annual Report to Department of Mines, 1899.
55. *Queensland Coals*, 6th ed.
56. Presumably John Petrie the eldest son of Andrew Petrie who founded a small engineering business in Brisbane in 1850.
57. *Moreton Bay Courier*, 29 November 1856.
58. Ibid., 30 May 1857; and 16 October 1858.
59. Ibid., 6 June 1857.
60. *Votes and Proceedings of the Legislative Council of New South Wales*, 1859–1860, Vol.1, p.1296.
61. *Statistics of Queensland*, 1860, p.53.
62. Ibid., 1862, p.52.
63. *Queensland Times*, 17 April 1860.
64. *Brisbane Courier*, 10 April 1862.
65. Queensland Government Titles Office (Old System), Special Grants 2C, Folio 11, 2 August 1867, Deed of Grant 17396; and Queensland Government Titles Office (Old System), Mortgage Deed to W.M. Boyd, 8 October 1859.
66. *Votes and Proceedings of the Legislative Council of New South Wales*, 1854, Vol.2, pp.1383–95.
67. Ibid.
68. Queensland Government Titles Office (Old System), Book 37, No.143, 3 November 1871.

The Second Mine at Redbank

5

The second mine at Redbank was opened on the south bank of the Brisbane River immediately to the north of the original Pioneer Mine and it was closely associated with one of the most colourful and forceful figures of early Queensland—John "Tinker" Campbell. Campbell had arrived in Sydney in 1833 with a consignment of tinware and clocks for sale but soon decided to try his luck in sheep farming and "a more shrewd, enterprising man never took up a clearing with more pluck and energy".[1] His venture prospered but in 1843 when the recession was depressing the price of wool, local pastoralists encouraged him to sell up both his stock and his property and invest the proceeds in Queensland's first boiling down works at Kangaroo Point.[2] Unfortunately he was unable to make the operation pay and after three years he sold out[3] and moved to Redbank where he collaborated with Joseph Fleming and Sir Alexander Stuart in their boiling down, timber and flour milling enterprises.[4]

By 1857 Campbell was approaching sixty years of age with an adult family and an eldest son, John Edwin, who showed an interest in entering the coal mining business. The timing was particularly opportune because a crisis in the supply of coal in the Moreton Bay area was once again developing. Of the two sources on which the district relied, the Moggill Mine was practically inoperative through flooding and in any case its reserves were almost exhausted; the life of Walter Gray's mine was severely limited by the small reserves of coal available to it and no moves had been made to supplement them. Campbell therefore looked for a suitable place to begin a new operation, selecting Redbank because it was in a convenient geo-

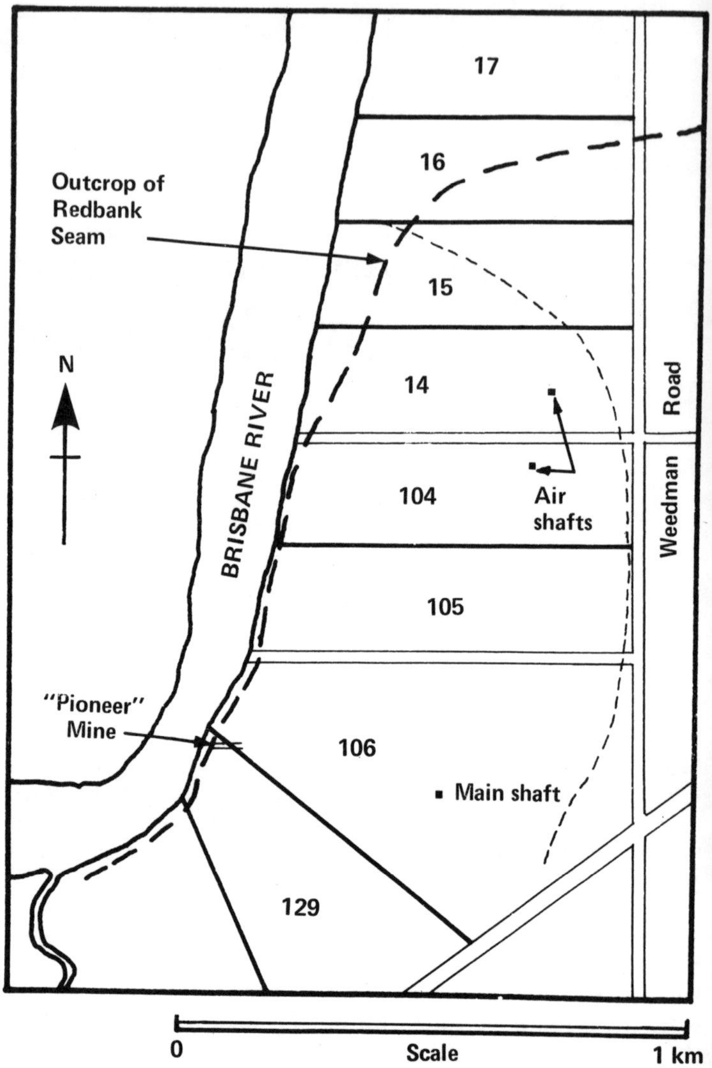

Figure 4. Redbank area showing subdivisions. Mine details are taken from Cameron's 1923 maps (chap. 5, n. 61). For location of the map see Frontispiece and also doc. 9, p. 97.

graphical position and adjacent to his existing interests.[5]

Eighteen months earlier under the stimulus of the colonial office policy of encouraging coal production in the Moreton Bay area without creating large monopoly companies,[6] a number of portions of land had been auctioned off at Redbank approximately one kilometre to the north of the Pioneer Mine where the coal outcropped on the Brisbane River.[7] Walter Gray had successfully bid £2 per acre ($10 per hectare) for Portion 16 Goodna and £1 per acre for the other two portions numbered 14 and 15 which adjoined Portion 16 to the south[8] and were nearer to Williams's original mine (figure 4). Gray was prepared to sell the three blocks, which had cost him £120 ($240) in total, for the substantial sum of £1,300[9] but this was beyond Campbell's means and he looked around for a partner to join his son and himself in their proposed undertaking. His choice fell on Robert Towns (plate 15).

Plate 15. Robert Towns. (Courtesy John Oxley Library.)

A greater difference in life-style and outlook than that between Campbell and Towns is hard to imagine. In contrast to Campbell's impetuous, flamboyant nature, centred on a love of club life, Towns was ". . . the incarnation of the virtues of thrift, sobriety, industry and perseverence".[10] He had risen to become a powerful figure in the commercial life of the colony through his business acumen and particularly by his relentless foreclosing on pastoralists who were in financial difficulties, although he often retained them to supervise the resumed properties as his managers.[11] He agreed to lend the Campbells £3,000 in return for a co-partnership which embraced their other interests " . . . of Fellmongering; of Boiling down Stock; and of Curing and Salting Meat and Hides".[12] The enterprise was given the name of John Campbell, Son and Another;[13] Messrs R. Towns and Company were appointed as its agents in Sydney[14] and J. M'Cabe carried out the same function in Brisbane.[15]

In the middle of the nineteenth century there was a choice between two methods for working a shallow, gently-sloping, outcropping coal seam such as occurred at Redbank. The first was to drive a tunnel or "day level" directly into the seam from the outcrop as had been done at the first Redbank Mine and at Moggill. Headings were turned off into the coal at regular intervals at right angles to the main drive and the coal worked "to the rise" between pairs of headings. The risk of the roof collapsing on the miners when the coal was extracted was minimized by dividing the working face into a series of "rooms" or "stalls", divided by "pillars" or "ribs", which took the weight of the roof as the coal was successively removed (document 7). Coal was always worked "to the rise" rather than "to the dip" because this ensured that water entering the mine through the strata could be drained away to a common sump instead of permitting it to collect in the advancing stalls where it inconvenienced the miners and was difficult to remove.

The alternative was to sink a vertical shaft to the seam well back from the outcrop and then drive to the rise. The coal could then be worked from headings turned off the main drive as before. The advantage of this method was that the underground movement of both the coal and the water was downhill to the foot of the vertical shaft where they could be hoisted mechanically to the surface. Additionally the oxidized (and usually poor quality) coal found adjacent to the outcrop could be left unmined if it was unsaleable. In the case of a seam outcropping on a river bank the elimination of openings in the outcrop had the additional advantage of reducing the risk of inundation during a flood. On the other hand the amount of development work required before production could begin was greater, and if the river were to be used for transporting the coal to the market a road or tramway had to be built from the shaft to the river. Redbank Mine was developed on an unsatisfactory compromise between the alternatives which included the worst features of both.

The Redbank Seam was up to 2.7 metres thick[16] and contained some 1.2 metres of workable coal,[17] the top 150 millimetres being interbanded with blackstone.[18] The outcrop commenced in Portion 14 Goodna some 3 to 6 metres above the river[19] falling with a dip of about 1 in 25 towards the south east[20] and reaching water level in that point to the pumping pit the System of working this Mine before the tunnel reached the Point at which the workings were being carried on (On the twentieth March One thousand eight hundred and sixty seven and during the whole of the said last mentioned time of five years) was to forewin the Coal that is to say the tunnel was driven forward One hundred and fifty feet a little more or less beyond the then last heading and from the face of the tunnel at that point to drive out headings East and West of the tunnel and from such headings to work the Coal by means of rooms driven to the North whereby the Coal was won from the dip to the rise the Coal Seam in the Redbank Mine dipping nearly South the System which I have above described is more particularly shewn in the following sketch

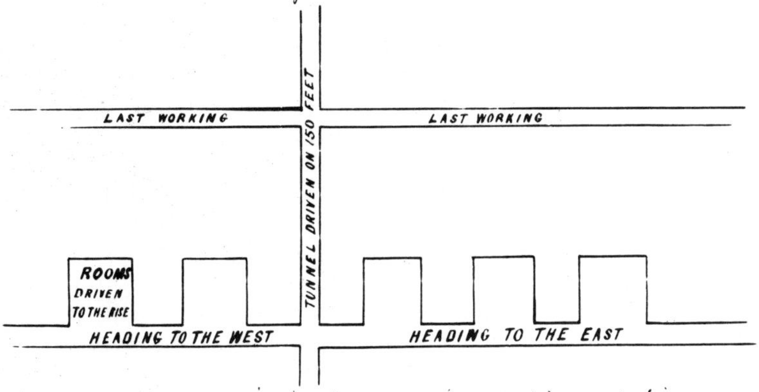

and during such last mentioned time of five years the Redbank mine was never worked on any other System than that which I have described until the tunnel arrived at the point to which the works were being carried on On the twentieth March One thousand eight hundred and sixty seven At this point the Mine having been worked on the System which I have described and all the Coal to the rise of the then headings having been won or nearly so and it having become necessary in accordance with the system of Working the Mine to drive forward the tunnel and throw out new headings I communicated

3 John Corbett

This is the third sheet of the affidavit of John Corbett sworn this fourteenth day of January in the year of our Lord One thousand eight hundred and sixty eight Before me [signature] A Commissioner for affidavits

Document 7. Extract from John Corbett's testimony in 1868 (chap. 5, n. 20) showing the method of working adopted at Redbank Mine. (Courtesy Queensland State Archives.)

Portion 129 (figure 4). The Campbells elected to sink a vertical shaft in Portion 14 but they located it only 400 metres from the outcrop (shaft A on document 8). This meant that in the area which they owned, a good deal of the seam was below the level of the foot of the shaft and consequently could not be worked to the rise.

Shaft sinking commenced in mid-1857.[21] The form of construction was typical of the time, the opening being oblong in section with dimensions of 10 feet by 5 feet (3 metres by 1.5 metres) and fully timbered for its 90 feet (27 metres) depth. At the foot of the shaft headings were driven east and west[22] and, for the first time in Queensland, a steam engine was installed on the surface to provide power both for the windlass which hoisted the coal[23] and to drive the pump located at the foot of the shaft. The pump bell-crank had a throw of 5 feet (1.5 metres) and rotated at 5 rpm. A vertical stalk suspended from the bell crank drove the pump which had a diameter of 5 inches (125 millimetres).[24] Redbank continued to be one of the very few coal mines in the district employing steam power until well into the 1870s. Permission was obtained to construct a wharf up to 100 feet (30 metres) long, jutting 10 feet (3 metres) out into the Brisbane River[25] in order to load the coal into punts or barges.

Although early samples had indicated that the coal might be suitable for gas-making, experience showed that it was much inferior to Newcastle coal in all respects.[26] Nevertheless early in 1859 the Campbells obtained the contract for supplying the vessels of the Australasian Steam Navigation Company,[27] but the shipping company soon became so dissatisfied with the quality of the coal that it moved the bunkering stop from Brisbane to Newcastle despite the inconvenience caused to passengers by the delay.[28]

Notwithstanding these problems some nineteen hundred tonnes of coal was raised in 1858[29] and a start was also made on boring deeper in the hope of finding a better seam. Late in 1859 an excellent sample was recovered from a depth of 140 feet (42 metres) below the workings.[30] The decision was immediately made to deepen the existing shaft so that it could be used for hoisting coal from the new seam.

Production from the existing seam could only be maintained by constructing another entrance to the mine and this was achieved by the simple—but short-sighted—expedient of intersecting the workings by a tunnel driven from the river bank adjacent to the coal wharf and commencing some 6 metres above the level of the water. Loaded tubs could then be hauled by horses direct to the loading shoots on the wharf along an underground tramway laid in the new drive[31] (document 8).

In 1860, portions 104 and 105 Goodna lying immedi-

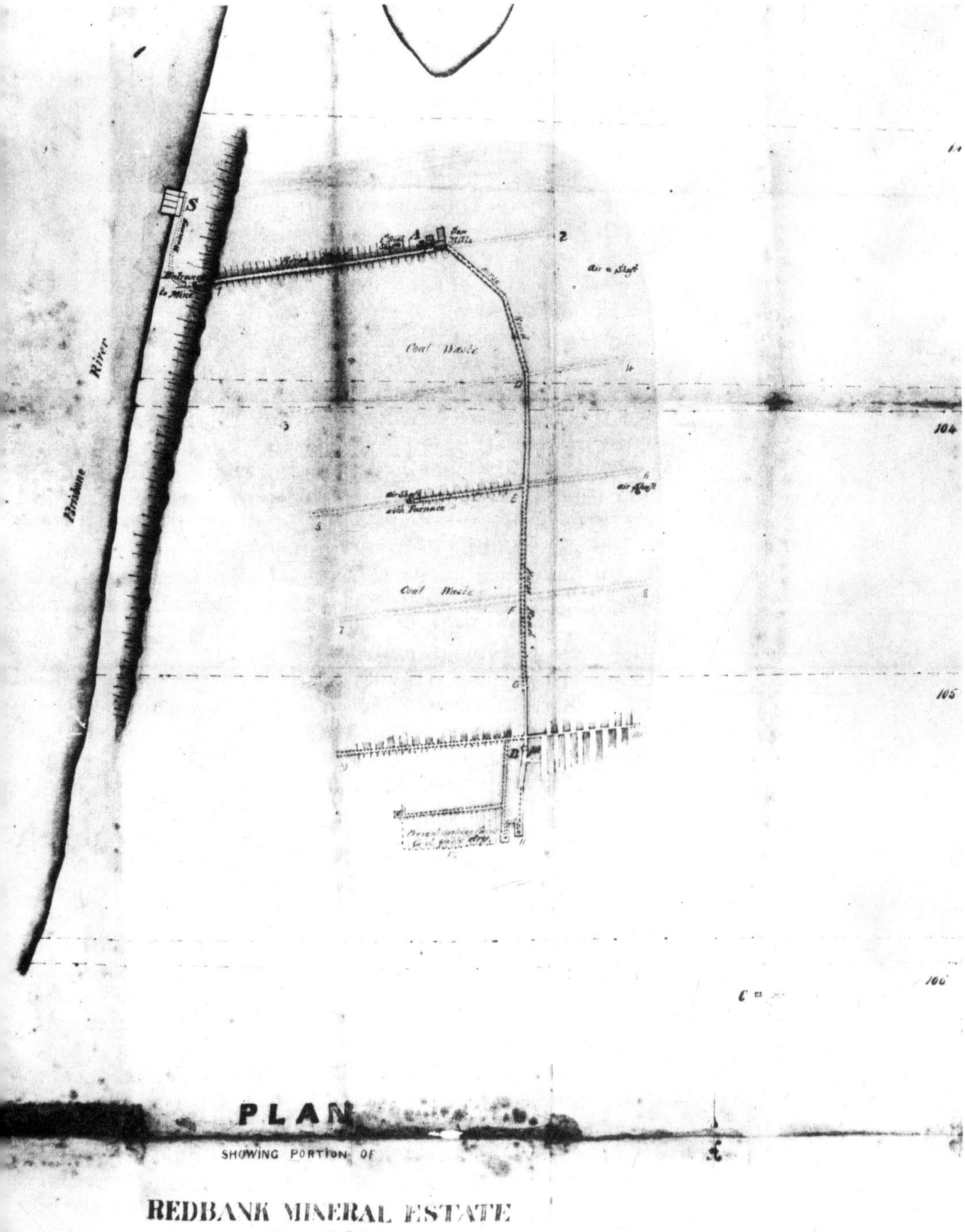

Document 8. Redbank Mineral Estate. Plan of underground workings prepared by James Robertson in 1868 (chap. 5, n. 17). (Courtesy Queensland State Archives.) See also fig. 4, p. 64, and doc. 9, p. 97, for other plans of the area.

ately to the south of the Redbank Mine were put up for sale by the first Queensland government as "Special Country Lots" being "supposed to contain workable coal"[32] (figure 4). The upset price was set at £2 an acre ($10 per hectare) and the Campbell and Towns partnership purchased them at that price.[33] The following year the Crown released portions 106 and 129 Goodna immediately to the south of Portion 105 Goodna and the partnership was again successful in its bids for them,[34] gaining a total area of workable coal of some 600 acres (240 hectares). These developments strained the already tenuous finances of the partnership and Robert Towns agreed to raise his stake from £3,000 to a holding of 50 per cent in the enterprise. A new deed was signed in July 1861,[35] and it was this document which was to precipitate a bitter legal battle between the signatories in 1867.

Meanwhile the partnership improved the quality of the coal as best it could, but chiefly by confining mining to the top metre of the seam and screening out the fines at the face.[36] The winding down of Walter Gray's mine towards its ultimate closure in 1862[37] deflected trade from his mine to Redbank because it was the only other regular producer at that time. By mid 1861 the quality of the product had improved sufficiently for the Australasian Steam Navigation Company to restart coaling its Sydney steamers in Brisbane.[38] New rolling-stock was purchased and improved loading facilities built including two 80 tonne bunkers on the river bank.[39] When the governor, Sir George Bowen, visited the operation in June he "... particularly expressed his surprise at the progress the coal mining works had attained under the management of the Campbells ... His Excellency also examined the mode of shipping coal, the engine and pump connected therewith ..." and concluded that they "... form the neucleus [sic] of what will hereafter prove the greatest coal mining district identified with the metropolis of Queensland".[40]

Unfortunately when the deepened shaft finally reached the new seam in October 1861[41] there was great disappointment. Instead of finding the 6 feet (2 metres) of coal which had been confidently predicted there were two beds each 2 feet (0.6 metre) thick separated by a layer of fireclay 18 inches (0.4 metre) thick.[42] A 2 tonne sample of the coal was shipped off to Britain for testing[43] but further work on the project had to be abandoned because the presence of the fireclay band made mining uneconomic even though the quality of the coal proved to be better than that of the higher seam.[44]

The earlier action of the owners in providing direct access by a tunnel to the main workings now left them with an idle and useless shaft because operations had by this time moved so far towards the south east that coal could no longer be mined to the rise from it. A sawmill was therefore constructed on the surface adjacent to the shaft and the ten horsepower steam engine used to power it[45] (document 8).

A major problem facing the partners at the end of 1862 was how the main workable seam could best be brought into full production in the area beneath the newly-acquired portions 104, 105 and 106 Goodna. The simplest plan was to turn the existing horseway southwards and continue mining to the dip, but a better and longer-term alternative was to sink a vertical shaft at a place on the new property where the seam was deepest. The second alternative was chosen and a site for a trial shaft selected near the north east corner of Portion 106 Goodna (shaft C on document 8). Sinking commenced early in 1863[46] but again the results were disappointing. At 118 feet (35 metres) a heavy inflow of water forced the cessation of operations, although exploration was pressed ahead for another 180 feet (54 metres) by boring an exploratory hole. According to the driller's log, no interception with the anticipated seams was made and it was therefore concluded that a slip or fault which had shifted the strata[47] was located somewhere between the existing workings and the trial shaft.

Meanwhile the grave disadvantages of having a tunnel entrance located on a river bank were brought home to the partners by severe floods in February 1863 and again in March 1864.[48] In 1863 the water was held in check by damming the tunnel entrance but in 1864 it swept all obstacles aside, inundating the mine and putting it out of operation for at least a fortnight.[49] A frame was later constructed which could be positioned quickly over the entrance in an emergency[50] but the obvious danger of intermittent flooding was always present. A suitable site for a new shaft entrance which would permit the permanent closure of the portal on the river bank was therefore sought.

The major surface feature of the newly-acquired land was a swamp which was near the centre of Portion 105 Goodna.[51] It was believed that a shallow cupping of the contours of the surface deposits was retaining the water in this area and producing the swamp. On the assumption that the coal measures formed the walls of the cup a shaft was commenced on the swamp in anticipation of intersecting the main seam at its lowest point in the locality (shaft B on document 8). The sinking of the 12 feet by 5 feet (3.6 metres by 1.5 metres) fully-timbered shaft commenced in 1863 or 1864 and the seam was reached at a depth of 90 feet (27 metres).[52] A 6 inch (150 millimetre) diameter reciprocating pump, operated from the surface by a 6 horsepower (4.5 kilowatt) steam engine was installed to handle the mine water[53] and the shaft used for downcast ventilation purposes. But despite its substantial cross-section it was never equipped for hoisting and the reason for this is unknown. The removal of foul air from the mine continued to be effected through unlined circular shafts which were sunk to meet the ends of selected headings. As was usual at that time a furnace at the foot of a ventilation shaft produced sufficient heat to draw fresh air through the workings and discharge it to the surface up the shaft.

Meanwhile mining continued to the rise from headings driven off each side of the horseway up which the coal was hauled to the shoots. The opportunity of using steam-powered rope haulage for this task had been lost in the early 1860s when the main access road had been turned through an angle of almost 90° to the south in order to work the coal under the southern portions. Because this horseway passed through mined-out areas and the roof was weak, its maintenance costs were high and the condition of the floor very poor.[54]

Nevertheless production rose steadily to about fifteen thousand tonnes per year during the period 1862 to 1865[55] but then it began to decline and by 1867 it was down to five thousand tonnes per year. To some extent the reduction in output was a result of the workings being advanced south of the new shaft to an area where the poor roof conditions and excessive drainage water made it impossible to maintain an adequate production face. In addition the horseway, located entirely in mined ground, was described as in "a disgraceful condition", ventilation needed improving by sinking at least one additional shaft and the pumping arrangements were inadequate.[56] In other words a major redevelopment was required.

The crisis came at a time when the partnership was in a state of disintegration. No extra finance was forthcoming and the Campbells were dragged through a lawsuit by Towns and forced out of the business.The mine struggled on for another two and a half years under a new manager, but in 1870 flood waters again interfered with operations, sweeping away the tramway and removing access to the shoots.[57] Operations were suspended despite the fact that the unworked coal was estimated (perhaps optimistically) at over twelve million tonnes[58]—or eight hundred years' reserves at the production rate ruling at that time! The Campbells retired to Macleay Island, turned their energies to growing sugar cane (which unfortunately was no more successful than their mining ventures),[59] and disappeared from the mining scene.

A fitting epitaph to the inadequate planning and poor management which brought about the premature closure of Redbank Mine was delivered by George Curless, a practical man with thirty-five years of coal mining experience in England and New South Wales who acted as overseer in its final days. At the hearing of the action brought by Robert Towns against the Campbells in 1867 for their mismanagement of the properties owned by the partnership he told the court:

> I have never seen any mine in all my experience in such a beastly state as the Redbank Mine. There appears to have been a system of mining adopted of getting the coal the easiest and best way for the time being. As the Redbank Mine stands at present no one with a knowledge of mining would invest capital in it.[60]

To-day most of the site of the Redbank Mine is covered by the store-sheds, workshops and sidings of Queensland Railways (plate 16). There is a small lantana-covered

Plate 16. Site of Redbank Mine, with Robertson's plan of the workings superimposed on it. (Photographed in May 1979.)

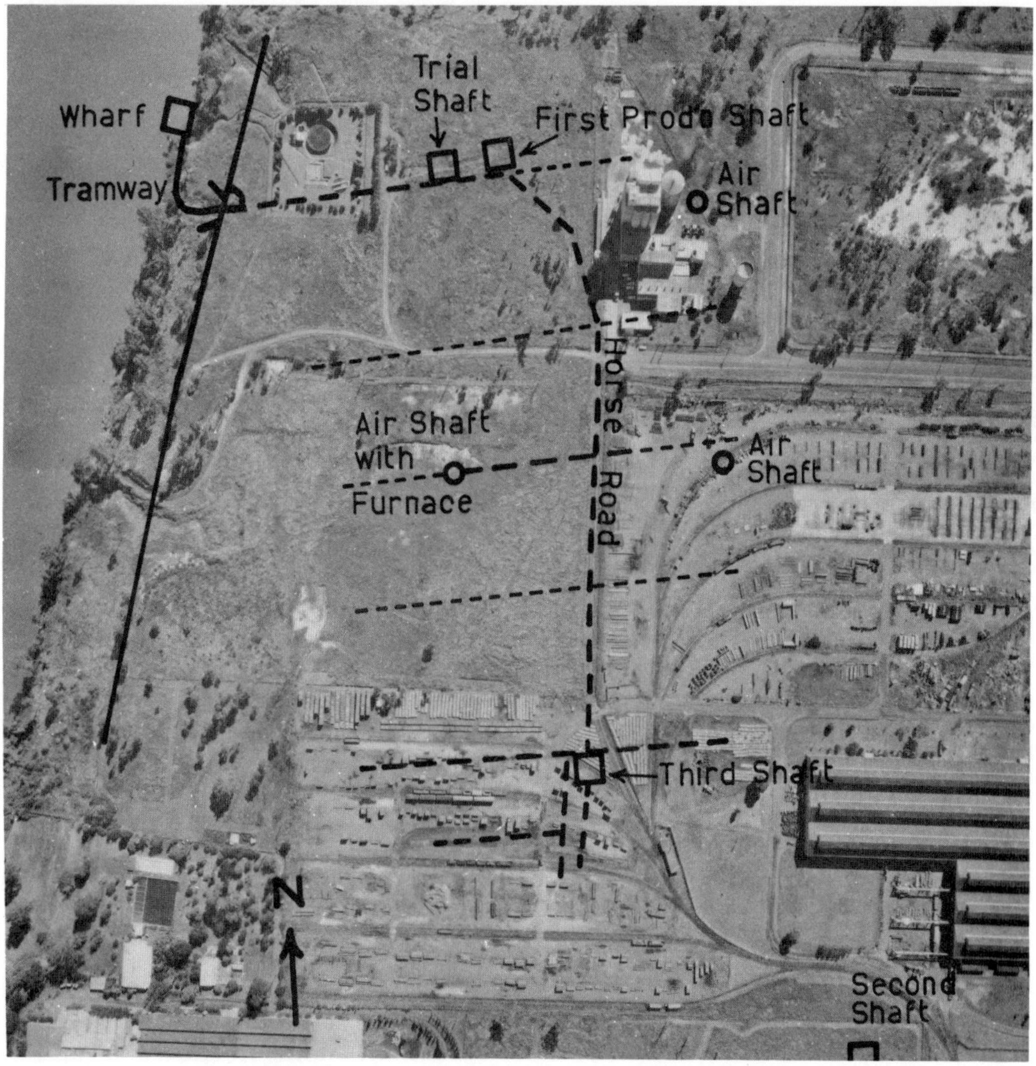

basin at the point where the horseway emerged onto the river bank and a circular depression nearby marks the position of a small shaft which was close to the portal (plate 17). The site of the original main shaft has not been built over and the surrounding ground has a high content of coal which may have come from one of the mine's spoil heaps. Any remains of the other two shafts and the various ventilation shafts have disappeared with the levelling of the ground which preceded the redevelopment of the area.

Plate 17. Redbank Mine. The tunnel entrance is marked by a white scar to the left of the first tree on the right. One of the shafts has left a circular depression in the ground behind the pair of trees immediately to the left of the entrance. The wharf was in line with the silos on the left. (Photographed in June 1980.)

Curiously, only two of the ventilation shafts and none of the major openings are marked on Cameron's 1923 map.[61] Denmead in his detailed report on the coalfield in 1955[62] seems to have been unaware of the existence of the second Redbank mine; he shows "Old Redbank Colliery" on Portion 105 but this reference is possibly to the mine which opened on Portion 106 many years after the closure of Campbell and Towns's mine and whose main shaft is included on Cameron's 1923 map (figure 4).

The Redbank Strike

Although there is a good deal of truth in the view that the genesis of trade unionism in mining in Australia can be found in the Eureka rising, the fact is that of the two mining unions which afterwards made up the strength of the Amalgamated Miners' Association it was the coal miners and not the gold miners who were the first to organize themselves.[63] Strikes had become potent weapons on the British coalfields in the 1850s and a large proportion of the miners entering Australia to work in the coal mines had had experience of them. In 1854, two whole years before the Eureka rising, the New South Wales miners in Newcastle had struck for a period of three weeks in order to win an advance in the hewing rate.[64]

By early 1861 the effect of the steadily falling price of coal on miners' wages led the union clubs in Newcastle to threaten strike action in support of a log of claims including "a proportionate day's wage for every operational miner" and an eight hour day.[65] After a series of inconclusive confrontations with the owners the Newcastle miners walked out as a body on 17 August 1861.[66] The industrial unrest and the pressures leading up to the strike inevitably filtered through to other coal mining communities in Australia, not least to those in the Moreton Bay region. Two months before the 1861 Newcastle strike, the new colony of Queensland experienced its first mining strike. It was at Redbank and was over a log of claims which had many elements in common with the demands being formulated by the Newcastle miners at that time.

The strike had its roots in a meeting held by the miners after work on Friday 7 June 1861, at which it was unanimously agreed that a deputation would call next day on John Campbell to present him with a series of demands supported by the threat of a strike on the following day if they were not met. The demands included a rise of 1s. per ton (10 cents per tonne) in the hewing rate (which at that time varied between 3s. 6d. and 6s. per ton), a specified week (which was usually in excess of ten hours per day at that time), and monthly pay settlements. Even if their demands were met the men proposed to reduce the daily output to a uniform two tonnes per man. The thrust of the proposals echoed those which the Newcastle miners had already drawn up in February of the same year, and a letter of support from their miners' association was read out at the Friday meeting.[67]

On the following afternoon a delegation of three miners, led by an Englishman, Edward Davis, approached John Campbell with the men's demands. Campbell refused point-blank to consider any of them and the delegation returned to the mine where the miners agreed that they would finish work for the day. Campbell saw the men leaving the pit and ordered the overseer, William Simpson, to go into the workings with some labourers, load the coal which had already been cut, and cart it to the shoots where a steamer was waiting to be coaled. Davis and one of the miners, William Griffith, saw the party disappearing underground and immediately went after them, demanding that they, "Bundle out of this". The labourers were obviously sympathetic to the miners and immediately left the mine where they were met by John Campbell and his second son, Frederick, who ordered them to return. By this time there were eight miners guarding the tunnel entrance and Davis threatened to put a pick through the head of the first man that went underground, and to shut up the mine unless their wage demands were met. There was a heated and threatening argument with Campbell, all work stopped and the eight miners were arrested for illegally conspiring to injure Campbell's interests by preventing anyone from working underground. The ensuing strike lasted for a period of about a fortnight.

In the police court the following Wednesday Mr Arthur Macalister (the newly-elected member of the Legislative Assembly for Ipswich and a future premier of Queensland), appeared for John Campbell. He made a particular point of stressing that it was a very important case both to workmen and employers of labour in the district, ". . . for experience had proved that combinations amongst workmen to raise the rate of wages had never proved beneficial to the workmen themselves". Whether he believed this remarkable statement to be true is unknown but the eight defendants were released on bail and remanded to the assizes where they were tried for conspiracy. However Campbell had overplayed his hand and the accused men were ably defended by John Malbon Thompson, a Sydney solicitor who had recently bought a practice in Ipswich[68] and had just been elected to the municipal council.[69] After some hesitancy the jury found all the defendants "not guilty" and they were released. It is not known whether the miners won the demands for which they struck but the short duration of the strike suggests that they did, or that an acceptable compromise was soon reached.

The militancy of the coal miners and their propensity to strike action was neither ignored nor forgotten by the government. In 1871, for example, when the Southern and Western Railway was considering whether to convert all its locomotives to coal firing the locomotive superintendent warned " . . . that alternative sources of coal should be maintained for railway locomotives".[70] The reason given for this recommendation was that "The miners, generally, are a troublesome class of men to deal with, and frequently strike". The result was that the coal contract was let to three suppliers on the basis of competitive tendering[71] and substantial stocks were held at key locations to insulate the railway from possible industrial action by the miners.

Notes and References for Chapter 5

1. *Australian Dictionary of Biography, 1788–1850*, Vol.1 (Carlton: Melbourne University Press, 1966), p.201.
2. *Pugh's Almanac* (Brisbane: T.P. Pugh, 1859), p.74.
3. *Moreton Bay Courier*, 16 January 1847.
4. F.W.S. Cumbrae-Stewart, "Some Queensland Memoir Writers", *Journal Historical Society of Queensland* 2 (1920): 1.
5. J. Campbell, *The Early Settlement of Queensland* (Ipswich: Ipswich Observer, 1875).
6. Commissioner Barney to Colonial Secretary, 23 September 1850 (Mitchell Library Microfilm A.227/629).
7. Queensland Government Department of Mapping and Surveying, Brisbane, Map M31.1.
8. Queensland Government Titles Office (Old System), Conveyance Book 6, Nos. 164, 165 and 176, 30 September 1856.
9. Queensland Government Titles Office, Transfer Book 1, No.170, 16 June 1857.
10. *Australian Dictionary of Biography, 1851–1890*, Vol.6, pp.294–96.
11. Ibid.
12. *North Australian*, 29 September 1857.
13. Queensland Government Titles Office, Mortgage Book 20, No.207, 28 August 1861.
14. *Towns* v. *J. and J.E. Campbell*, 1867. Queensland State Archives. Accession 25/90. SCT.36 (Alexander Stuart's Testimony).
15. *Moreton Bay Courier*, 25 July 1857.
16. *North Australian*, 15 March 1859.
17. *Towns* v. *J. and J.E. Campbell*, 1867.Accession 25/92. SCT.38 (James Robertson's Testimony).
18. Ibid., Accession 25/90.SCT.36 (Thomas Moody's Testimony).
19. *Brisbane Courier*, 23 October 1861.
20. *Towns* v. *J. and J.E. Campbell*, 1867. Accession 25/92. SCT.38 (John Corbett's Testimony).
21. *Moreton Bay Courier*, 13 February 1858.
22. *Towns* v. *J. and J.E. Campbell*, 1867. Accession 25/92. SCT.38 (Robertson).
23. *Moreton Bay Courier*, 13 February 1858.
24. *Towns* v. *J. and J.E. Campbell*, 1867. Accession 25/92. SCT.38 (Robertson).
25. *Moreton Bay Courier*, 3 November 1858.
26. Ibid., 16 May 1857; and 13 March 1860.
27. *North Australian*, 15 March 1859.
28. *Moreton Bay Courier*, 1 March 1860.
29. *Statistics of New South Wales*, 1858.
30. *Brisbane Courier*, 23 October 1861.
31. *Towns* v. *J. and J.E. Campbell*, 1867. Accession 25/91. SCT.37 (George Curless's Testimony); and Accession 25/92. SCT.38 (Corbett).
32. Queensland Government Department of Mapping and Surveying, Map M31.1.
33. Queensland Government Titles Office, Vol.2a, Folio 71 and 72, 22 November 1860.
34. Ibid., Vol.3a, Folio 113 and 128, 27 March 1861.
35. *Towns* v. *J. and J.E. Campbell*, 1867. Accession 25/90. SCT.36 (Stuart); and Accession 25/90. SCT.36 (John Campbell's Testimony).
36. Ibid., Accession 25/90. SCT.36 (Moody); and Accession 25/92. SCT.38 (Corbett).
37. *Statistics of Queensland*, 1862.
38. *Moreton Bay Courier*, 16 May 1861.
39. *Towns* v. *J. and J.E. Campbell*, 1867. Accession 25/91. SCT.37 (Curless); and Accession 25/92. SCT.38 (Thomas Francis's Testimony).

40. *Moreton Bay Courier*, 5 June 1861.
41. *Brisbane Courier*, 22 October 1861.
42. *Towns* v. *J. and J.E. Campbell*, 1867. Accession 25/92. SCT.38 (Robertson).
43. *Brisbane Courier*, 28 November 1861.
44. *Towns* v. *J. and J.E. Campbell*, 1867. Accession 25/91. SCT.37 (Curless).
45. Ibid., Accession 25/92. SCT.38 (Robertson).
46. Ibid.
47. Ibid.
48. *Brisbane Courier*, 17 February 1863; and 22 March 1864.
49. Ibid., 17 February 1863; and 28 March 1864.
50. *Towns* v. *J. and J.E. Campbell*, 1867. Accession 25/91. SCT.37 (Curless).
51. Ibid., Accession 25/90. SCT.36 (Moody).
52. Ibid.
53. Ibid., Accession 25/92. SCT.38 (Robertson).
54. Ibid., Accession 25/91. SCT.37 (Curless).
55. *Statistics of Queensland* 1861-1868, incl.
56. *Towns* v. *J. and J.E.Campbell*, 1867. Accession 25/91. SCT.37 (Curless); Accession 25/92. SCT.38 (Robertson); and Accession 25/91. SCT.37 (Curless).
57. *Queenslander*, 4 December 1869; and 26 March 1870.
58. *Towns* v. *J. and J.E. Campbell*, 1867. Accession 25/92.SCT.38 (Robertson).
59. *Queensland Government Gazette*, 1870, Vol.11, p.536.
60. *Towns* v. *J. and J.E. Campbell*, 1867. Accession 25/91. SCT.37 (Curless).
61. W.E. Cameron, "Geological Map of Ipswich and Bundamba Coalfields", *Queensland Geological Survey*, Publication 271 (New Series, Vol.8), Sheet 3, Map 2 (Brisbane: Government Printer, 1923).
62. A.K. Denmead, "The West Moreton (Ipswich) Coalfield", *Queensland Geological Survey*, Publication 279 (Brisbane: Government Printer, 1955).
63. B. Fitzpatrick, *The British Empire in Australia* (Carlton: Melbourne University Press, 1941), p.285.
64. M.H. Ellis, *A Saga of Coal* (Sydney: Angus & Robertson, 1969), p.42.
65. Ibid., p.72.
66. Fitzpatrick, *British Empire in Australia*, p.287.
67. *Moreton Bay Courier*, 14 June 1861; and 9 August 1861.
68. *North Australian*, 29 December 1857.
69. *Australian Dictionary of Biography*, 1851-1890, Vol.6, pp.265-66.
70. *Votes and Proceedings of the Legislative Assembly of Queensland*, 1871, pp.661-73.
71. *Queensland Government Gazette*, 1872, Vol.13, p.572.

Other Operations on the Brisbane River

6

During the 1840s and 1850s it is probably true that all the obvious outcrops on the Brisbane River were visited and examined with a view to starting commercial production but only two of them in addition to those already mentioned were significant enough to be worthy of notice.

In his 1854 survey,[1] Stutchbury had referred to a coal outcrop on the north bank of the river, approximately two and a half kilometres south west of Williams's Moggill Mine. He was impressed by the situation and commented that "This place may be considered as one of the few exceptions in favour of 'day level' working as there is a fine creek and easy access to the river, the main coal being ten or more feet thick, situate sixty feet above the river's edge". In 1853 Thomas Prior had purchased land nearby in the Pocket which now bears his name[2] and shortly after Stutchbury's visit he is believed to have opened a mine on the spot to which reference was made.[3] No contemporary information on its operation has so far come to light; it figures in none of the annual statistics and it may be assumed that any contribution which it may have made to local production was very small.

In order to identify the other area of activity it is necessary again to return to Stutchbury's thirteenth report.[4] He began his tour from Brisbane and opened the report by saying:

> I again visited the [Moreton] bay, and on my way down, stopped at the "Stone Quarries", situate in the parish of Bulimba, about eight miles down the Brisbane River, on the south side, where I had before observed coal in the banks, upon land belonging to Mr. Charles Coxen [sic]. The coal pre-

> sents a weathered surface of nearly ten feet in thickness . . ."

The importance of the position of the deposit was not lost on Stutchbury who continued

> . . . its dip being towards the river, and about twenty feet above high water, gives it, if it should upon driving into it turn out good, a great advantage as regards its shipment, and the slope of the sandstone below being at the moderate angle of six degrees, presents no difficulty in the formation of wharves. From its position being so near to Moreton Bay, its presence must, in a considerable degree, advance the value of land in this neighbourhood, as compared with that upon which coal is now but slightly worked thirty miles higher up the river.

(This last comment was obviously a reference to the Moggill Mine.)

At the time, Charles Coxon, a well-known pastoralist with extensive leases on the Darling Downs,[5] owned portions 18 and 19 Bulimba on the river bank (figure 5) and immediately following Stutchbury's departure a trial pit was excavated on his land by a small syndicate. "Accordingly an opening was made close to the river side, and so fair are the prospects of success, that the four proprietors are likely to be able to supply coal to shipping lying alongside their pits . . . ".[6] After a number of vicissitudes, including dissolution of the original partnership, some "excellent coal" was procured and a sample put on board the packet steamer *Boomerang* for use in tests on her return trip to Sydney.[7] Altogether some two hundred tonnes of coal were mined,[8] but the results were disappointing. Twenty years later John Sinclair, who was by then a partner in Smellie's foundry in Brisbane, commented "I think I was the only engineer who used it; you could scarcely call it coal, as it was black stone. We had on one occasion to stop and throw about 20 tons of it overboard, we could not get steam out of it".[9] As on so many other occasions the bright appearance of the Moreton Bay coal had belied its high ash content and relatively poor quality which made it difficult to market.

In 1859 the press reported that a drift had been run into the hillside on Charles Coxon's land and a shaft sunk nearby, but without finding any better coal.[10] The possibility of reopening the original workings was reassessed from time to time,[11] but the quality of the coal did not justify the effort involved. Both workings were still visible forty years later, at which time the government geologist, Mr Robert Jack, found that the seam was only 0.6 metre thick and confirmed that the coal was of poor quality.[12]

John Williams, while still heavily engaged in his operations at Moggill, had purchased Portion 15 Bulimba adjacent to Bulimba Creek on 22 July 1853,[13] but he does not appear to have prospected it seriously until 1862

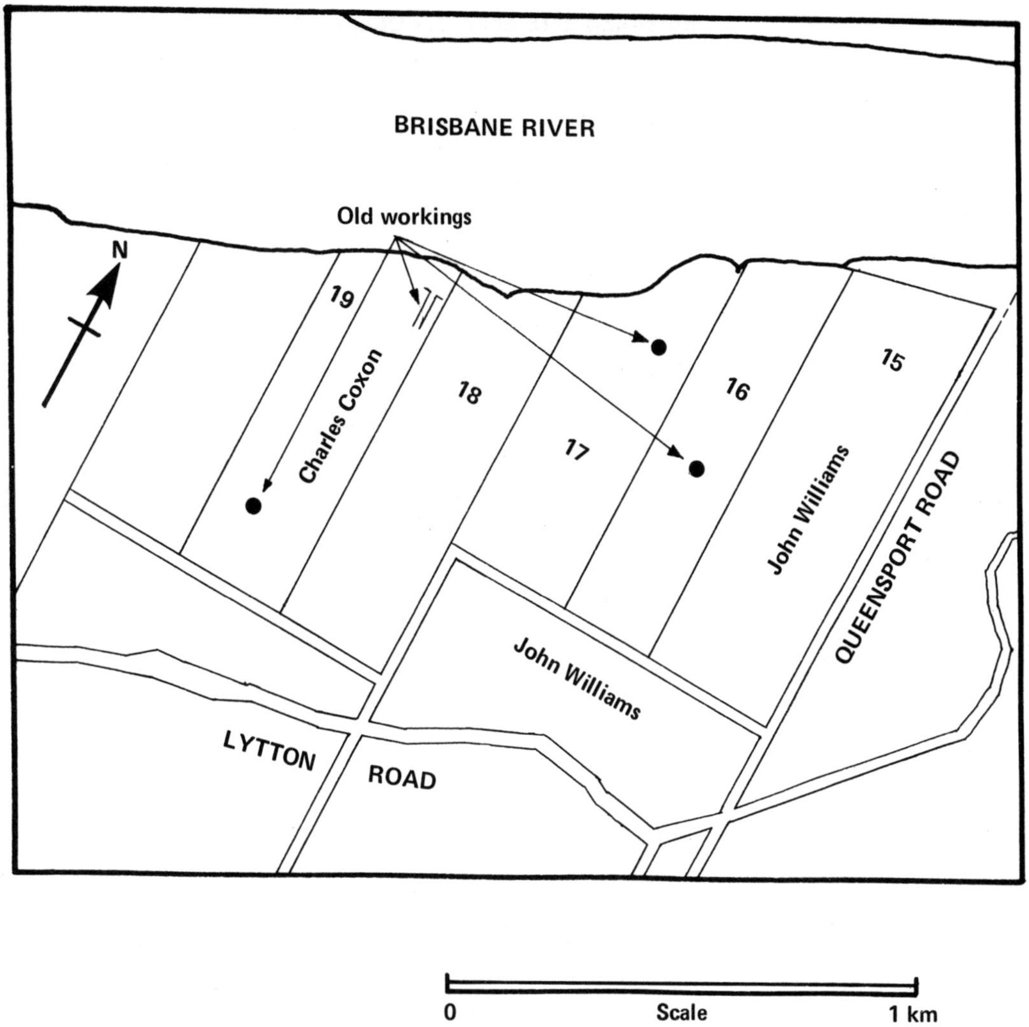

Figure 5. Old workings on the lower Brisbane River. (After E. O. Marks, "Coal Measures of S. E. Moreton", *Queensland Government Survey*, Publication 225, 1910.) See Frontispiece for location of the map.

when a shaft was sunk which intercepted coal, but it was not considered sufficiently encouraging to justify a mining operation.[14] Exploration work continued intermittently in the area for many years but little, if any, coal warranting the opening of a mine ever seems to have been found. In later years the prospecting methods employed came in for severe criticism[15] but the hard fact ultimately had to be faced that although the location of Bulimba was ideal from the transport point of view, the quality of the coal in the area was too poor for it to compete on the home or overseas market. Subsequent development of the area has obliterated all surface indications of earlier mining activities.

Notes and References for Chapter 6

1. *Votes and Proceedings of the Legislative Council of New South Wales*, 1854, Vol.2, pp.1383–95.
2. Queensland Government Titles Office (Old System), Conveyance Book 2, Nos. 200, 201, 202, 203, 204, 205, 206, 30 March 1853.
3. L.C. Ball, "Prior's Pit", *Queensland Government Mining Journal* 14 (1913): 301–2.
4. *Votes and Proceedings of the Legislative Council of New South Wales*, 1854.
5. *Australian Dictionary of Biography*, 1851–1890, Vol.3 (Carlton: Melbourne University Press, 1969), pp.487–8.
6. *Moreton Bay Courier*, 24 June 1854.
7. Ibid., 20 October 1855.
8. *Brisbane Courier*, 18 November 1863.
9. *Votes and Proceedings of the Legislative Assembly of Queensland*, 1878, Vol.2, pp.319–402.
10. *Moreton Bay Courier*, 16 February 1859.
11. *Brisbane Courier*, 18 November 1863.
12. Annual Progress Report for 1895, *Queensland Geological Survey*, Publication 108 (Brisbane: Government Printer, 1896), pp.4–6.
13. Queensland Government Titles Office (Old System), Conveyance Book 2, Nos. 200, 201, 202, 203, 204, 205, 206, 30 March 1853.
14. Annual Progress Report for 1896–1898, *Queensland Geological Survey*, Publication 143 (Brisbane: Government Printer, 1899) pp.6–11.
15. Ibid.

Queenland's Early Years

7

Separation Day, 10 December 1859, marked the beginning of a particularly favourable period for the rapid settlement and development of the new colony of Queensland. The demand for Australian wool which had been strong in the 1850s continued to strengthen in the 1860s as Britain and Europe raced into industrialization.[1] From 1861 to 1865 the American civil war severed the Lancashire cotton mills from their suppliers of raw material in the southern states, compelling the textile manufacturers to give increased attention to wool and its industrial products.[2] By 1870 Australia was providing about 70 per cent of the wool imports to the United Kingdom as compared with only 40 per cent of a very much smaller import in 1860[3] and immigrants poured into the country, hoping for a grant of land and the opportunity to establish themselves as graziers. They found that this was not as easy as they had imagined.

The diversion of labour to gold mining in New South Wales and Victoria in the 1850s had made it impossible for farmers to find enough shepherds to tend their flocks. An immense effort had gone into fencing the runs,[4] stabilizing the number of sheep, raising their quality and improving their yield of wool.[5] By the beginning of the 1860s a situation had been reached in the southern colonies where there were fewer sheep than at the beginning of the 1850s[6] but the opportunities for settling profitable areas had greatly diminished because of the fencing which had taken place. On the other hand in Queensland, which had hardly been touched by the gold rushes, there was plenty of land suitable for sheep farming. Consequently immigrants flooded into the new colony both from the southern colonies and directly from overseas.[7] In a decade the population quadrupled to 115,000 and at its peak was increasing by over 20 per cent per annum.[8] The number of sheep practically trebled and wool production increased sevenfold to a plateau where it remained with minor changes until the 1880s.[9]

The strong, steady overseas demand for wool ensured that the selling price to which it had slowly climbed during the 1850s was maintained in the 1860s and when the price temporarily eased at the end of the decade economic stability was fortunately maintained by a rapid rise in the export of gold. Over the same period there was a general fall in import and domestic prices, resulting in a rise in the real value of money.[10] Despite government inexperience in financial management, solid, sustained growth in all areas of activity was therefore possible, with receipts and expenditure more than quadrupling during the new colony's first decade.[11]

The pastoralists established themselves in a broad belt parallel to the coast where the rainfall was most reliable, the eastward-flowing rivers forming natural routes for the supply of the necessities of life to the settlers and the transport of wool down to the sea. Here the coastal steamers carried it to Sydney or Brisbane for shipping to Europe, although at a later date direct shipments began from some of the northern ports of Queensland. In response to the expansion and the scattering of the population there was a startling improvement in the number and quality of roads, together with the appearance of specialist haulage contractors. The cost of carriage by road was considerably reduced, falling by up to 50 per cent between 1857 and 1864 in New South Wales and showing the same trend in Queensland a few years later.[12]

The major inland river port in Queensland was Ipswich (plate 18) although ships from the southern colonies were unable to reach it because of the tortuous course of the Brisbane River and its shallow depth.[13] The problem was particularly acute at Seventeen Mile Rocks until the obstruction was removed by explosives in 1866[14] and for many years it was usual for most goods to be transshipped at Brisbane. A regular daily steamer service between Brisbane and Ipswich was provided by the Australasian Steam Navigation Company[15] but in addition there was a variety of side-paddle,[16] stern-paddle[17] and screw-propeller vessels[18] plying for trade on a charter or private basis. During the 1860s the steamboats on the Brisbane River multiplied so quickly that by 1870 there was treble the 1860 number.[19] At the same time coastal steamer services were increasing in frequency, and larger, more comfortable boats were being brought into use. The Sydney service became twice weekly instead of twice

Plate 18. Ipswich in 1860. (Courtesy John Oxley Library.)

monthly and there were regular weekly sailings to Rockhampton and Gladstone.[20]

The growth of the wool industry was followed by the expansion of other primary industries and the establishment of new ones. The traditional boiling down of cattle was supplemented by the establishment of a meat processing industry.[21] Sugar cane became a regular crop,[22] while the American civil war led to a short but dramatic rise in cotton cultivation.[23]

Simultaneously, support industries began appearing in the centres of population. In 1862 Alexander Cameron opened the first foundry in Brisbane.[24] Steam engines became common sources of power for wool pressing, cotton ginning,[25] ice manufacture,[26] flour milling,[27] brewing,[28] sugar crushing[29] and refining[30] and water supply;[31] Brisbane gas works was opened in 1865[32] and the statistics show that in 1869 there were 115 stationary steam engines in use in the colony in addition to 60 mechanically-operated plants of various kinds.[33]

Gold production was one of the few activities which was not stimulated by Separation and it actually declined during the early years as the alluvial deposits at Canoona were worked out. The finding of gold at Clermont in 1862, followed by discoveries at Thane's Creek and Calliope in 1863 reversed this trend, but it was the rush to Gympie in 1867 which made the greatest impression on the fortunes of the colony because it stimulated gold prospecting in the north. In the following year gold was found in the Gilbert Ranges, quickly followed by discoveries on the Etheridge

River and in a number of other districts. The economic result of the steep rise in gold production which began in the late 1860s was to compensate for a weakening in the wool market. The lure of gold drew hundreds of unemployed from Brisbane into the rest of the colony and during the four years 1865 to 1869, gold output rose almost twelvefold, compared with that of the preceding five years. The total value of mineral production increased from one sixth of the wool clip in 1865 to almost a half by 1870, the lion's share being contributed by gold.[34]

Coal Mining

For a variety of reasons the coal mining industry did not enjoy the degree of expansion which might have been expected from the dynamic rise in activity in the colony. Unfortunately the statistics for coal production throughout the 1860s are scanty and those which are available must be treated with caution.As late as 1869 the government statistician complained that "At present the amount [of coal] raised is comparatively small, but . . . the quantity raised is not reported, and as its consumption is almost confined to the Colony, the Export Returns do not supply information on the subject".[35] The most reliable figures are probably those which were assembled in 1871 by a commission reporting on the extension of the railway from Ipswich to Brisbane.[36] The commissioners concluded that at the end of the decade the coal consumption of seagoing steamers had stabilized at about twelve thousand tonnes per annum, with some two thousand tonnes per annum being used by the river steamers and two to three thousand per annum going for domestic purposes and general industry. The increase from five thousand tonnes per annum in 1859 to seventeen thousand tonnes per annum in 1870 may seem modest in the light of the great expansion in other parts of the colony's economy but there were sound reasons for it.

In the first place the increase in steamer traffic was to some extent balanced by a marked improvement in the efficiency of coal utilization. The low-pressure side-lever marine engines of the 1840s had been replaced by grasshopper, oscillating, or steeple engines using high-pressure steam which, if not superheated in the modern sense of the word, was at least dry.[37] The rectangular boilers had been replaced by fire-tube types which gave greatly improved heat exchange, while jet condensers had given way to surface condensers which permitted the boilers to run on recirculated fresh water instead of sea water.[38] Whereas the 75 kilowatt side-lever engines of the *Shamrock* consumed coal at a rate of some 4.8 kg/kW/hour,[39] the *Diamantina* which came into service in 1861 (and was of a comparable size and horsepower) had a fuel consumption of only one half of this figure[40] (plate 19). Thus the quadrupling of the steamer service to Sydney barely doubled the coal consumption, even though the ships were larger; the trebling in the horsepower of the boats on the Brisbane River was achieved on barely twice the coal consumption.

Plate 19. The steamship *Diamantina*. (Courtesy I. J. Farquhar, Dunedin.)

Secondly, the boilers which raised the steam for the engines which in increasing numbers powered the industries and services of the new colony were only fuelled by coal when its price compared favourably with wood,[41] and there was still a plentiful supply of timber adjacent to the population centres until the early 1870s, when for the first time it became more expensive than coal.[42] Thirdly, coal exports which at one time had been viewed very optimistically[43] had failed to materialize because of the relatively poor quality and high friability of the coal and the inadequate loading facilities at the port of Brisbane.[44]

The decade had started well enough. In 1861 the Australasian Steam Navigation Company reversed its previous decision and recommenced the regular coaling of its steamers in Brisbane despite intermittent problems with the quality and supply of the coal.[45] Campbell and Towns who were the main suppliers from their Redbank Mine erected a depôt adjacent to the Australasian Steam Navigation Company's stores in order to provide facilities for all steamships using the port.[46] In this optimistic atmosphere samples of Redbank and New South Wales coals[47] were sent to England for assessing their suitability for naval use,[48] and although the Redbank coal was not recommended a large block of it formed the centrepiece of the Queensland exhibit for the 1862 International Exhibition in London.[49] During the Newcastle strikes of 1862[50] export sales of almost four thousand tonnes were achieved from Brisbane[51] and the government statistician noted with a certain optimism that " . . . coal mining is gradually increasing and will no doubt before long become a very important occupation".[52]

In 1865 the Queensland government revived a dream of the previous decade by sponsoring a regular mail service between Australia and Singapore via the Torres Strait.[53] First the *Soucheys* and then the *Hero* were chartered to make the run between Brisbane and Batavia, and on its third (and final) trip the owners of the *Hero* courageously fuelled their ship entirely with Ipswich coal from the Old Tivoli Pit.[54] Although the failure of this attempt to establish the route could not be blamed completely on the quality of the coal there is no doubt that its high friability contributed to its inadequate steaming properties and thus to the delay of the *Hero* by almost a week on its return journey.[55]

After 1863 coal exports dwindled practically to nothing as a result of intense competition from the Newcastle mines.[56] The sharp fall in its selling price could not be matched by the Queensland mines and even led to a steady import both of coal and coke into the colony.[57] By the early 1870s imports had climbed to more than six thousand tonnes per annum,[58] the chief consumer being the Brisbane gas works which was unable to find a satisfactory local alternative supply.

The coming of railways to Queensland in 1865 made little immediate impact on the production or transport of coal because the locomotives were initially fired with wood (plate 20). The first public line from Ipswich to Toowoomba and Warwick intersected the Walloon coal measures at Rosewood and other smaller deposits along its route, but the only ones which were immediately exploited were at Allora and Clifton near Warwick where limited production began in 1870.[59]

When the railway was extended from Ipswich to Brisbane in the mid 1870s there was considerable discussion regarding the extent to which it would replace the river as the main means of transport for passengers and raw materials,[60] but no attempt was made to connect it to any of the mines even though, for example, the track crossed the land being mined at Goodna. Nor did the government make any move to encourage the export of coal by constructing coaling wharves or handling facilities in Brisbane as the New South Wales government had already done at Newcastle. Coal continued to be held in the

lighters in which it had been carried down river; when it was required it was loaded onto the ships by hand.[61] Only in the early 1880s was a serious attempt made to use the railways for handling coal between the Ipswich mines and the Brisbane or overseas markets.

Plate 20. Queensland's first railway locomotive on a trial run between Ipswich workshops and the railway wharf, 11 February 1865. (Courtesy John Oxley Library.)

The railways were, however, inadvertently instrumental in opening up a small export market for Queensland coke. Before finally deciding to fuel its locomotives with coal in 1870, the railway had let out a contract to the Old Tivoli Pit for the supply of coke for its engines and the owners—Hooper and Robinson—had commenced operating a battery of coking ovens to supply it.[62] When the contract was withdrawn the owners received compensation of £900[63] and used the surplus capacity to establish a small export trade in coke with Victoria and South Australia.[64]

The modest rise in coal production which was achieved in the decade following Separation was accompanied by a steady fall in the selling price. From 15s. per ton ($1.50 per tonne) at the pit mouth in 1859[65] it had dropped to 9s. per ton by 1871, only rising during the 1861 Newcastle coal miners' strike when on one occasion it actually reached 16s. per ton.[66] The dwindling profit margin depressed the general state of the industry and discouraged new entrants into it. The total number of coal miners in 1864 was 74 and this figure was not exceeded until 1871 when it climbed to 122.[67]

Coal Producing Areas

For most of the decade following Separation production was centred on Redbank and Goodna, although after 1866 coal mining restarted near Ipswich because the coals were of a higher quality and included coking varieties.[68] There was also a brief return to the Moggill area where pits were opened for a short time in the 1850s or 1860s,[69] in addition to the already mentioned Darling Downs pits a few years later.

The chief producing mines were on the banks of the Brisbane River which had become the main coal transportation artery in the colony, particularly after the navigation had been improved by blasting a passage through the notorious Seventeen Mile Rocks, to which reference has already been made.[70] Attempts to open mines at any distance from the Brisbane or Bremer rivers created transport problems for their owners who generally had to cart the coal along inadequate roads to riverside shoots[71] because of the financial and legislative difficulties involved in constructing private tramways.[72]

Outside the Brisbane area coal was discovered north west of Rockhampton in 1862,[73] in Laidley[74] and at Burrum[75] in 1863, at Blair Athol in 1864[76] and on the Bowen River in 1867[77] (plate 21). A cannel coal suitable for kerosene production was found in Bungil Creek, 50 kilometres north of Roma in 1865,[78] and there were regular reports of similar discoveries in the next decade. However, the only one of these areas from which there was any coal production before 1870 was Burrum[79] and this was small and intermittent despite the coal being tested on the steamers

Plate 21. Richard Daintree's drawing of the coal seams at Pelican Creek on the Bowen River. (Reprinted from R. Daintree, *Queensland, Australia*. London: G. Street, 1873.)

Queensland and *Clarence*[80] and pronounced as better than the Redbank product and the equal of Newcastle coal.

After Separation a doubling of coal production was achieved every five years, but it must be remembered that the 1859 base figure of five thousand tonnes per annum from which it grew was very small, particularly when judged by the New South Wales figures. For example, in 1860 the coal output from the Newcastle area alone was one third of a million tonnes. Tests which the British Admiralty carried out on it in 1862 were very favourable and it quickly gained a high reputation overseas for quality. Despite repeated industrial unrest the annual New South Wales production passed half a million tonnes in 1864 and by 1870 had reached three quarters of a million tonnes of which one third was exported.Railway connections were built between the mines and the port at Newcastle where the government constructed coal loading facilities which handled in a single week the equivalent of the total production of Queensland for a whole year.[81]

The Queensland mines suffered from a number of handicaps compared with their southern neighbours. The disadvantages of higher ash, lower specific energy and greater friability were aggravated by working the coal near to outcrops where deterioration due to oxidation was most in evidence.[82] This method of working had come about to some extent by the land management policies adopted by the colonial legislature in the 1840s. A policy had been followed in which "All deeds of grant . . . should henceforth convey to the purchaser everything below, and everything above the surface",[83] and the more obvious or convenient coal-bearing areas in the Moreton Bay region had been divided into tiny allotments of less than two hectares for auctioning at an upset price considerably in excess of the normal price of £1 per acre ($5 per hectare) for agricultural land. In contrast the original Australian Agricultural Company in Newcastle had had an almost unlimited area available for coal mining when it began operating in 1825. As late as 1859 the Newcastle Wallsend Coal Company was able to purchase a single 500 acre (200 hectare) estate to start its operations, with the option of purchasing a further 1,039 acres from the government (at 5 guineas per acre, i.e. $26.25 per hectare) if required.[84] It might have been better if possession of the most valuable known portions of the coalfields in the Moreton Bay area had been retained by the New South Wales government until the resources of the district had been better understood and were ready for development,[85] even if illicit mining on Crown Lands had continued in the meantime.

One of the first actions of the new government in 1860 was to approve the Alienation of Crown Lands Act[86] which gave the governor the power to sell at his discretion lands for mining purposes without competition, and at a price of £1 per acre ($5 per hectare), provided that the area did not exceed 640 acres (256 hectares). He also had the power to issue conditional grants of land for mining minerals. In these cases, 2s. per acre was generally deposited and if at the expiration of one year it could be shown that spending at the rate of £1 per acre had taken place for mining purposes the selector was allowed to purchase the land at an upset price of £1 per acre without competition. But coal and gold were specifically excluded from the terms of the Act.

Regulations for the conditional granting and non-competitive purchase of crown lands containing coal were not introduced until 1865[87] when the MLA for Maryborough, William H. Walsh, urged the premier to act.[88] The regulations demanded a deposit of 5s. per acre and if, within one year from the date of the letter of approval of the application, the selector could show that he had "mined and conveyed to some place where it is available for steam vessels or manufacturing purposes, a quantity of good merchantable coal, not being less than one ton for each acre of land selected by him" and pay the remaining 15s. per acre he was entitled to the land in fee simple. However, if these conditions were not met the land and all the works on it reverted to the Crown and the deposit was forfeited. The area which could be leased in this way was not less than 80 acres nor more than 640 acres and the legislation formalized arrangements which had evolved over the preceding years.[89] The Burrum River coalfield was the first to be affected by the new regulations and shortly afterwards they permitted J.M. Thompson to acquire for £100 the 100 acre Portion 270 Goodna[90] at Blackstone which for many years was the main source of coal for the highly profitable Aberdare Mine.

There was sharp criticism at the time of the introduction of the legislation that coal was being treated less favourably than other minerals and a plea was made for the differentiation to be removed.[91] The 1867 amendments to the Alienation of Crown Lands Act of 1860, which were consolidated in the 1868 Act,[92] achieved this but at the expense of raising the deposit on all mining lands to 5s. per acre and making the balance payable within twelve months, subject to satisfactory evidence of having spent £1 per acre on it. At the same time the minimum area which could be sold under these terms was reduced to 40 acres, the maximum remaining at 640 acres. In 1872 the minimum and maximum areas were reduced to 20 and 320 acres respectively resulting in the creation of tiny portions which later had to be combined into workable properties on such sites as New Chum and Old Aberdare mines.

The first coal mining enterprise in Queensland which was able to amass an adequate tract of land was at Redbank and it resulted from the partnership between John Campbell, his son, and Robert Towns, to which reference has already been made. By 1861 they had assembled about 500 acres (200 hectares) by purchases from private owners and from the government[93] and were able to carry out a drilling programme to assess the reserves.

The pattern of small, shallow, short-lived, privately-owned coal mines run by individuals or partnerships became a feature of the district which persisted for over a century. In contrast the trend in Newcastle was towards the mines being owned by public companies which, if successful, could raise sufficient finance to maintain at least a modest development and equipment programme. Most Queensland operations were run on a "shoe string". For example, the surface arrangements at Old Tivoli Pit were described as being ". . . such as are usually found about coal mines of the least possible extent and pretentions".[94]

Whether stronger government initiatives towards encouraging the establishment of larger mines would have been desirable or successful in the years following Separation is questionable. The major obstacles to higher production were the indifferent quality of the coal, the poor returns offered in mining it, the difficulty of transporting it from sites not situated immediately on the Brisbane or Bremer rivers and the intense competition for overseas markets from New South Wales. These factors far outweighed any restraints resulting from unsympathetic or unsuitable legislation.

Notes and References for Chapter 7

1. B. Fitzpatrick, *The British Empire in Australia* (Carlton: Melbourne University Press, 1941), pp.192–98.
2. Ibid., p.189.
3. A. Barnard, *The Australian Wool Market 1840–1900* (Carlton: Melbourne University Press on behalf of Australian National University, 1958), p.218.
4. Fitzpatrick, *British Empire in Australia*, pp.192–98.
5. Barnard, *Australian Wool Market*, p.15.
6. Fitzpatrick, *British Empire in Australia*, p.177.
7. Barnard, *Australian Wool Market*, p.11.
8. *Statistics of Queensland*, 1860 and 1870.
9. Statistical Handbook of the Sheep and Wool Industry, 3rd ed. (Canberra: Bureau of Agricultural Economics, 1961), p.40.
10. Barnard, *Australian Wool Market*, p.9.
11. *Statistics of Queensland*, 1860 and 1870.
12. Barnard, *Australian Wool Market*, pp.11, 84, 182–83.
13. *Votes and Proceedings of the Legislative Council of New South Wales*, 1855, Vol.2, p.1117.
14. *Votes and Proceedings of the Legislative Assembly of Queensland*, 1866, p.31.
15. *Moreton Bay Courier*, 11 June 1861.

16. R.H. Parsons, *Paddle Steamers of Australasia* (Adelaide: Private Publication, 1967).
17. *Brisbane Courier*, 18 July 1863.
18. Ibid., 21 February 1862.
19. *Statistics of Queensland*, 1860 and *Votes and Proceedings of the Legislative Assembly of Queensland*, 1871, pp.675-837.
20. *Brisbane Courier*, 18 July 1863 and 30 April 1863.
21. *Towns* v. *J. and J.E. Campbell*, 1867. Accession 25/90. SCT.36 (Alexander Stuart's Testimony).
22. N.L. McKellar, *From Derby Round to Burketown* (St. Lucia: University of Queensland Press, 1977), p.35.
23. Fitzpatrick, *British Empire in Australia*, p.189.
24. *Brisbane Courier*, 19 July 1862.
25. Ibid., 14 April 1862.
26. Ibid., 6 February 1865.
27. Ibid., 14 October 1861 and 15 November 1862.
28. Ibid., 20 August 1863.
29. *Queenslander*, 6 April 1867.
30. *Queensland Times*, 18 February 1871.
31. *Brisbane Courier*, 18 July 1862.
32. Ibid., 30 November 1865.
33. *Statistics of Queensland*, 1869.
34. W. Fryar, "The Development and Progress of Mining and Geology in Queensland", *Proceedings Australian Association for the Advancement of Science* 4 (1895): 361.
35. *Statistics of Queensland*, 1869.
36. *Votes and Proceedings of the Legislative Assembly of Queensland*, 1871, pp.675-837.
37. *Brisbane Courier*, 7 March 1865; 16 October 1861; 19 April 1865; and 15 October 1861.
38. Ibid., 19 April 1865.
39. *Votes and Proceedings of the Legislative Council of New South Wales, 1852*, Vol.2, pp.1007-1039.
40. *Queenslander*, 2 February 1867. When first built the *Diamantina* was claimed to have a coal consumption of less than 1 kg/kW/hr but this figure was later found to be much too optimistic.
41. *Brisbane Courier*, 4 November 1865 and 8 December 1865.
42. *Queensland Times*, 28 October 1871.
43. *Brisbane Courier*, 23 October 1861.
44. *Votes and Proceedings of the Legislative Assembly of Queensland*, 1871, pp.675-837.
45. *Moreton Bay Courier*, 30 April 1861.
46. *Brisbane Courier*, 26 July 1862.
47. M.H. Ellis, *A Saga of Coal* (Sydney: Angus & Robertson, 1969), p.63.
48. *Brisbane Courier*, 28 November 1861.
49. Ibid., 30 October 1861.
50. Ellis, *A Saga of Coal*, p.84.
51. *Statistics of Queensland*, 1862.
52. Ibid.
53. *Votes and Proceedings of the Legislative Assembly of Queensland*, 1865, pp.949-1031.
54. *Queenslander*, 22 December 1866.
55. Ibid., 16 February 1867.
56. Ellis, *A Saga of Coal*, pp.93-100.
57. *Brisbane Courier*, 6 February 1865; 4 November 1865; 27 June 1863; and 16 February 1864.
58. *Statistics of Queensland*, 1873.
59. Ibid., 1870.
60. *Votes and Proceedings of the Legislative Assembly of Queensland*, 1871, pp.675-837.
61. Ibid.
62. Ibid., pp.661-73.
63. Ibid., 1871-1872, p.31.

64. Ibid., 1871, pp.675-837.
65. *Moreton Bay Courier*, 2 April 1859.
66. Ibid., 5 September 1861.
67. *Votes and Proceedings of the Legislative Assembly of Queensland, 1864*. Second Census, 1 January 1864, pp.625-48; and 1872, Fourth Census, 1 January 1871, pp.965-1262.
68. Ibid., 1871, pp.675-837.
69. A.K. Denmead, "The West Moreton (Ipswich) Coalfield", *Queensland Geological Survey*, Publication 279 (Brisbane: Government Printer, 1955).
70. *Votes and Proceedings of the Legislative Assembly of Queensland*, 1871, pp.675-837; and 1866, pp. 31-32.
71. *Brisbane Courier*, 6 April 1867.
72. *Votes and Proceedings of the Legislative Assembly of Queensland*, 1878, Vol.2, pp.319-402.
73. *Brisbane Courier*, 29 September 1862.
74. Ibid., 15 October 1863.
75. *Maryborough Chronicle*, 3 October 1865.
76. A.W. Jose and H.J. Carter, *The Australian Encyclopaedia* (Sydney: Angus & Robertson, 1925), pp.274-77.
77. *Queenslander*, 16 February 1867.
78. *Brisbane Courier*, 4 November 1865.
79. *Votes and Proceedings of the Legislative Assembly of Queensland*, 1866, p.1473.
80. *Brisbane Courier*, 29 August 1865; and 28 July 1865.
81. Ellis, *A Saga of Coal*, pp.93-100, 63, 60.
82. *Brisbane Courier*, 16 September 1865.
83. British Parliamentary Papers: Australia 1840-41 (Shannon: Irish University Press, 1970), p.171.
84. Ellis, *A Saga of Coal*, pp.23, 60.
85. *Brisbane Courier*, 24 October 1861.
86. Queensland Act, 24° VIC, No.15, 17 September 1860.
87. *Queensland Government Gazette*, 1865, Vol.6, p.746.
88. *Queensland Parliamentary Debates*, 1865, Vol.3, pp.464-74.
89. *Votes and Proceedings of the Legislative Assembly of Queensland*, 1865, pp.1207-8.
90. *Queensland Times*, 20 December 1866.
91. *Brisbane Courier*, 21 September 1865.
92. Queensland Act, 31° VIC, No.46, 28 February 1868.
93. *Towns* v. *J. and J.E. Campbell*, 1867. Accession 25/90. SCT.36 (Stuart).
94. *Brisbane Courier*, 6 April 1867.

Goodna Mine

8

By 1860 small rural communities had established themselves along the banks of the Brisbane River, particularly at points on the south bank where the main turnpike road connecting Ipswich to Brisbane approached the river. One of these settlements had been officially recognized in 1856 and named Goodna.[1] With the digging of foundations and sinking of wells to satisfy the needs of the growing community the surface geology of the river plains became better understood and it was realized that Woogaroo Creek marked the edge of a large coal-bearing region which stretched in a fairly continuous sheet westwards towards Ipswich.[2]

A short distance to the west of the boundary of Goodna township and about one hundred metres south of the main road to Ipswich[3] a coal seam outcropped in the side of a small hill formed by a fold in the strata. In 1860 the Queensland government divided this area into twenty-six small farms[4] and offered them for sale. Portion 109 Goodna which ran down to the main road from the summit of the hill was purchased by David Jenkins in 1861[5] and came to be known locally as Jenkins's Paddock[6] (document 9).

During this time a few kilometres further to the west at Redbank, the Campbell and Towns partnership was developing its boiling down, curing, fellmongering and coal mining interests.[7] The quality of the coal was causing concern and a satisfactory fuel could only be produced from a small section of the seam.[8] The partnership hoped that the quality of the coal might improve nearer Goodna and with this in mind it purchased Jenkins's Paddock in June 1865[9] after discovering that it contained a coal seam

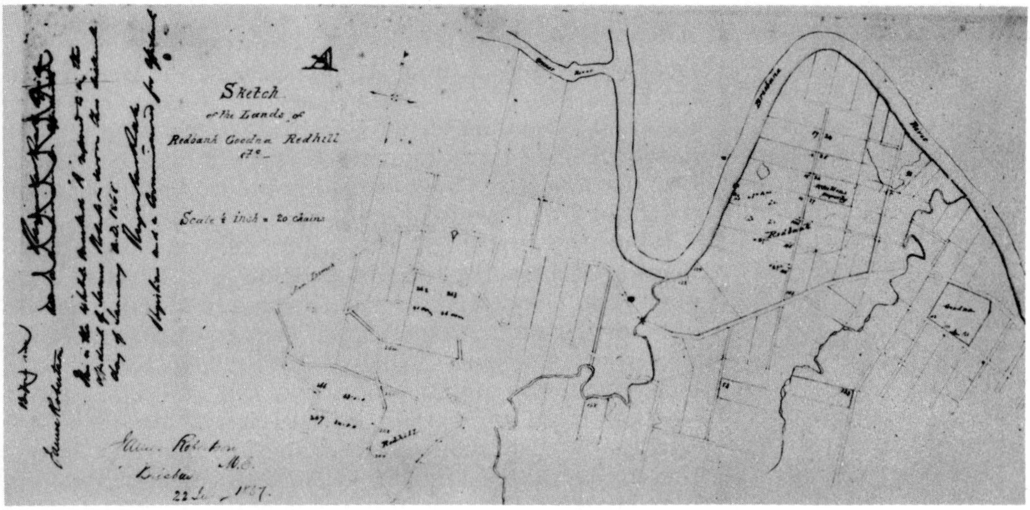

Document 9. Goodna, Redbank and Redhill mines. Plan prepared by James Robertson in 1867 (chap. 8, n. 3).

about 4 feet 6 inches (1.35 metres) thick with a gentle dip of 3° to 5° towards the south.[10]

Mr John Corbett, a solid, honest and practical man with thirty years' experience as a coal miner in Scotland and New South Wales was overseer for the Campbell and Towns's Redbank Mine in 1865 when the purchase was made, and the partners entrusted him with opening the mine at Goodna. He tackled the job by driving a tunnel directly into the outcrop and working the coal to the rise from headings turned off the main drive. After only a few months a fault was encountered and the quality of the coal deteriorated. Corbett pushed ahead hoping that the faulted area was of limited extent but conditions did not improve and after advancing a further fifty or sixty metres he decided to abandon the tunnel and sink a shaft instead.[11]

A preliminary borehole had indicated that the coal improved in quality as the seam dipped gently away to the south and the shaft was therefore sunk close to the southern boundary of the land and about 900 metres from the road.[12] Corbett chose a very small cross-section of only 7 feet by 4 feet (2.1 metres by 1.2 metres) for the shaft which was centrally divided by a brattice to provide both a downcast and an upcast section despite the flow of air in them being seriously impeded by the presence of the tiny balanced cages each measuring only 3 feet 8 inches by 2 feet 8 inches (1.1 metres by 0.8 metres).[13]

Arrangements underground were a little unconventional. The shaft was oriented with its long axis running east-west perpendicular to the dip but the first two headings were not driven east and west as was usual practice at the time. Instead, a single heading 5½ metres wide was driven north to the rise for about 10 metres and divided down the centre by a timber and brattice-cloth partition. Headings were then turned off to the east and west resulting in the loss of a 10 metre panel of coal near the foot of the shaft.[14] Goodna turned out to be a fairly wet pit[15] and it was claimed that this method of working was adopted because it assisted in channelling the water to the foot of the shaft for subsequent disposal.[16]

When the rupture in the Campbell and Towns partnership occurred in 1867 production had only just commenced at Goodna Mine and Towns's first action was to appoint the highly experienced George Curless as overseer in place of John Corbett.[17] Output was stepped up to compensate for falling production from the older pit at Redbank,[18] and the area to the rise lying between the shaft and the outcrop was soon worked out. In 1869 Towns therefore acquired a further 140 acres (56 hectares) of land immediately to the south of the existing workings under the terms of the mineral clause of the Alienation of Crown Lands Act of 1868.[19] At the same time Thomas P. Moody, a well known and highly respected mining engineer from New South Wales who had appeared for Towns in the 1867 litigation case against the Campbells, was made superintendent of the whole Redbank estate enterprise which included the Goodna Mine.[20] Moody constructed a tramway to the river,[21] workings were developed to the south of the shaft and the ventilation system was reorganized.[22] By 1870 production had reached eight thousand tonnes per year, a good deal of which was sold to the Australasian Steam Navigation Company and a little to the Southern and Western Railway.[23] A vivid description of the operation was given by a reporter who visited the mine in 1871.

> This mine, the property of Messrs. R. Towns & Co., Redbank estate, is situated about fifteen chains from the boundary of the Goodna township in a westerly direction. It is connected with the river, which is only half-a-mile distant, by a wooden tramway, which leads to a large shoot divided into three compartments, capable of holding 280 tons. The mine is not to be seen from the Brisbane road, but the tramway, and the appearance of coal, cinders, &c., scattered about the place, are sufficient indications to lead a visitor in the direction of the works. The first building which attracts the eye is the residence of the superintendent of the mine, Mr. Wm. [sic] Curless, which is within sight of the road. On the ridge behind the house, a short distance off, the mine is situated. The first thing to attract notice is a building 34 ft. by 14 ft. used as stables, and to which there is a store-room attached. A little apart from these is the engine-room, 18 ft. by 22 ft., covered in with a galvanised iron roof. The engine is a six horse-power portable, working on the second motion. The head-gear of the pit is the next thing to be observed. It comprises two wire ropes, an inch and an eighth in thickness, attached to the drum of the engine, on which the cages containing the waggons of coal hauled from the pit are slung; the pulleys on

which these wire ropes run are five and a half feet in diameter, and stand thirty feet above the platform. The dimensions of the shaft are seven feet by four feet, and there is a partition in the centre which divides the shaft into two equal parts. A gauge runs down either side of these partitions in the centre to keep the cages steady when the coal is being hauled to the top. The load raised amounts to 15 cwt., including weight of waggon and cage; and as the load comes up one division, the empty waggon goes down the opposite. A signal-wire is fixed on the right hand division, and communicates with a hammer in the engine-house, certain strokes indicating to the engine-driver what is wanted by the man in charge of the loading. At the top of the shaft there are "catches" fixed the entire way across, which are self-acting by means of a wooden spring, so that when the cages pass up the shaft they push these catches back, and when they pass beyond them the spring shoots them out again to a distance of about three inches; the cages resting on the ledge thus formed until the pit headsman removes the full waggon, and replaces it with an empty one. He then draws back a lever which removes the catch, and the cage again descends. Both sides of the pit work in the same manner simultaneously, and a great deal of labour and time are saved by the ingenuity of the plan.

The depth of the pit shaft is eighty-five feet, with what is known in mining as a "sump" on one side only, seven feet deep. Into this sump all the water which percolates through the coal is conducted; and the water-standish [sic], or well, is adequate to contain over 500,000 gallons. So constant is the dripping of the water through the coal that to prevent its accumulation a man and a boy are constantly engaged in bailing it into waggons, which have to be wheeled up-hill, a distance of 220 yards, to the sump already referred to. About forty waggons per day are thus removed, and the work has to be kept up constantly the whole year round. For the final removal of this water from the sump an iron water-tub is employed, the working of which strikes the eye of a visitor at once as being peculiarly ingenious, while at the same time it is very simple. The manager had it put in operation while we were present, so that we could see it in all its parts. It is three feet in depth and two feet three inches square, and is completely self-acting. On descending to the bottom of the pit, it loads itself in the sump by means of a valve in the bottom four inches square, which yields to the pressure of the water as the weight of the tub carries it down, closing when the tub becomes full; and on returning to the top of the pit again it discharges itself by means of a lever connected with a similar valve on the side of the tub, near the bottom, the lever being acted upon by a block of wood fixed on the side of the framing above the pit. The appliance is one which suits the purpose remarkably well, and is a wonderful improvement on the old plan of removing the water by means of waggons hauled to the top in the same manner as the coal.[24] It is the invention of Mr. Curless, the superintendent, as are also the catches already referred to, and only very recently adopted; the saving of time and labour thus effected is very considerable.

These works were all that were to be seen from the top, and, having obtained one of the miner's little lamps, we descended the shaft in company with Mr. Curless, by means of one of the cages for conveying coal-waggons. It was some time, however, before we could see anything, and not until we had become a little accustomed to the blackness and darkness of the place. A door having been opened, we had a further look into the sump, and heard the water trickle into it in little streams. We then proceeded to inspect the pit itself and its workings, proceeding through the main drive. There is not the least appearance of fire-damp in any part of the mine, nor has there ever been any known to exist since its opening. At the bottom of the pit there are two levels or headings, which have been driven east and west, six feet wide by four and a half feet high. From these all the coal has been completely worked out to the rise of the pit. It is the portion of the mine commenced in Jenkins's paddock, thirty-three acres having been purchased in the first place by Campbell, Son, and Another, and which have now been in the possession of Messrs. Towns & Co. for over six years. The coal from this has been completely exhausted, that which remains being of no marketable value.

At present there is a main drive opened which is carried due south to the dip of the coal, a distance of 220 yards, sixteen feet wide, by four and a half feet high. A double tramway is laid down the entire way to facilitate the removal of the coal to the main shaft, the miners wheeling all their own coal, with the assistance of "helpers-up" provided by the proprietors of the mine. This is pretty hard work, as the dip shows a gradient averaging two inches in the yard. From the main drive there are six headings opened—three to the east and three to the west—at regular distances apart of about two and a half chains. The various rooms are turned off from these headings at a distance of twenty-nine feet apart, nine feet being allowed to a pillar and twenty feet to a room, the pillars to be worked out at some future time. The full thickness of the seam of coal is twenty-three feet throughout the mine, of which four and a half feet only is workable, the remainder being of a very inferior quality. This four and a half feet of coal is intersected by six bands of fire-clay, which are taken out and thrown into the "gob"—the empty space left behind the miners as they progress onwards. There are thirty-one miners engaged, each of whom restricts himself to eight waggons per day, the entire quantity raised being sixty to seventy tons per day. In driving headings the men are paid four shillings per ton, and three shillings for every yard driven, the entire width being six feet only; in the rooms they are paid at the rate of four shillings per ton, without any extras. The roof of the mine is of coal, and is a good substantial roof, nineteen and a half feet through; the bottom is composed of black shell. A seam of coal two feet ten inches thick underlies the present seam at a depth of four feet and a half, but it is scarcely of so good a quality as the coal in the present seam.

The propping of the mine is regular and should be very durable. The props are erected about a yard apart in two rows each side of the tramway into the face of the mine, and four rows are erected in the road leading into the sides. The entire workings of the mine show great regularity and neatness, and bespeak the most careful as well as skilful management.

The ventilation of the pit has been admirably arranged. An air-shaft or "in-take" as it is termed, has been sunk to the west, a distance of 200 yards, and to the depth of thirty feet; it has been sunk on the crop of the coal to save additional sinking. The air going down this in-take travels from west down towards the south, traversing each room by an air-way carried along the sides, the several rooms being connected by "cut throughs" made to allow the air to traverse the face of every room. The air is conveyed from one heading to another by doors placed in such positions as the manager's experience tells him are the most suitable. From the south the air passes on towards the east, and thence to the north, where a large furnace has been erected built of brick four feet wide, five feet nine inches high, and twelve feet long, and connected with the coal-shaft at a distance of twenty-three feet from the bottom. A fire is kept constantly burning, when the pit is in working, for the purpose of creating artificial ventilation, and the one side of the coal-shaft itself is used as the "up-cast" or out-get for the air.

The land upon which the mine is situated, containing about 140 acres, was taken up under the mineral clause of the Land Act of 1868, by the Hon. R. Towns some eighteen months ago, so that all the extensive workings of the pit described have been executed within that short period of time. Mr. Curless is the sole manager of everything in connection with the mine, and the skill with which every portion of the workings has been executed proves him to be a man of very large experience in coal mining.

It only remains to add that the coal is removed from the pit on trucks made to hold four waggons of five hundred weight each, drawn by two horses on the wooden tramway described above, the driving to the shoot being done by contract by a man and a boy; a relay of horses relieves those put on in the morning at dinner-time. The tramway is of a two-foot gauge laid on sleepers, which are notched, and each rail is fitted into a notch and secured by a wedge.

The coal is of a very excellent quality, and the principal consumers are the A.S.N. Co., the S. and W. Railway, Messrs. J. and G. Harris, Messrs. Bright, Brothers, and Co., the Government steamers, the steam dredge and tug in Brisbane, as well as a large household trade in Brisbane.

Owing to the increased expenses of driving water and coal uphill in the present mine, the proprietors have come to the conclusion to open the mine to the south of the present workings, so that the coal may be worked to the rise. Sinking a new shaft has therefore been commenced about twenty-six chains from the present pit. The shaft is being opened nine feet by five feet inside, and the probable depth will be 120 feet. The surface layer is composed of nineteen feet of red clay, underneath which there is rough granular sandstone. The sinking of the shaft has been let by contract, the proprietors finding all the necessary plant, as well as all the erections, including a horse-gin, head-gear, &c. The drum of the gin is nine feet in diameter. This gin will be used in removing all the surface shift, [sic] etc., and it is intended, if the coal proves of good quality, to erect a fifteen horse-power engine capable of hauling 200 tons of coal per day, as well as of keeping the mine dry by means of five-inch pumps.

> The manager, Mr. Curless, was very courteous in pointing out everything in connection with the mine, and in giving every information regarding it. At some future time we hope to report upon the new workings just now commenced.[25]

The coal from Goodna required cleaning[26] because its quality was poorer than that from the newer pits being developed at Tivoli and Bundamba,[27] but production continued to be maintained in the early 1870s at about seven thousand tonnes per year.[28] The new shaft was apparently never completed beyond the exploratory stage, and the problem of hauling the coal underground up dip to the foot of the shaft was eased by introducing horses underground.[29] With the death of Towns in 1873[30] and the return of Thomas Moody to New South Wales,[31] interest in the mine faded and it did not compete for the railway contract in 1874 despite the fact that the new Ipswich to Brisbane line ran within twenty metres of the pithead. Production fell away and the mine was put up for sale in 1876 but apparently with no buyers.[32] It disappeared from the government statistics in 1877,[33] closure probably being due to a lack of demand caused by its inability to compete with the higher quality coals coming from the newer Ipswich mines.[34]

Although the site of Goodna Mine has been completely obliterated by subsequent mining operations and the area now stands derelict, two possible connections with the original enterprise can be identified. There is a likelihood that the overseer's house, which had four rooms and a verandah,[35] still stands in its original position on Portion 112, although the present building was extensively repaired and modified around 1970 (plate 22), except for

Plate 22. George Curless's house at Goodna. The house has been extensively modified at various times but the roof is believed to be original. (Photographed in May 1980.)

Plate 23. The tramway decline at Goodna showing the bottom of the ramp leading from the mine area to the river where the wharf was located. (Photographed in May 1980.)

the roof whose design and materials of construction are appropriate to the 1870 period. The second connection relates to the tramway from the mine to the Brisbane River. After crossing the Goodna to Ipswich road it descended the bank to the river plain, and a cutting through the bank can still be identified which lines up exactly with the position of the main shaft, as given by Robertson in his 1867 plan (document 9; plates 23, 24). Denmead[36] marks this earthwork as the site of an old tunnel but a careful inspection indicates that it is in fact a decline.

The trial shaft which in 1871 was reported as being sunk 26 chains (520 metres) from the original site in order to open up the reserves to the south of the workings was presumably located on Portion 241 which Towns had purchased in 1869[37] and which is now separated from the mine site by the Ipswich to Brisbane railway line and the Cunningham Highway. The few remains which can be identified on Portion 241 probably belong to Goodna No. 1 Mine, which operated some forty years after Towns's pit had closed. As they are less than 400 metres from his original shaft it is unlikely that they are located on the site of the second shaft whose exact location therefore remains unknown.

Plate 24. Site of Goodna Mine, showing George Curless's house and (as a broken line) the possible course of the tramway down the ramp to the river. (Photographed in May 1979.) The position of the mine was taken from James Robertson's 1867 map (chap. 8, n. 3). See document 9, p. 97.

Notes and References for Chapter 8

1. *North Australian*, 29 January 1856.
2. *Votes and Proceedings of the Legislative Assembly of Queensland*, 1871, pp.675-837.
3. *Towns* v. *J. and J.E. Campbell*, 1867. Queensland State Archives. Accession 25/92. SCT.38 (James Robertson's Testimony).
4. Queensland Government Department of Mapping and Surveying, Brisbane, Map M31.22.
5. Queensland Government Titles Office (Old System), Deed 2440, 26 March 1861.
6. *Queensland Times*, 28 October 1871.
7. Queensland Government Titles Office, Mortgage Book 20, No.207, 28 August 1861.
8. *Towns* v. *J. and J.E. Campbell*, 1867. Accession 25/92. SCT.38 (Robertson).
9. Queensland Government Titles Office, Vol.30. Folio 225, 226. 13 June 1865.
10. *Towns* v. *J. and J.E. Campbell*, 1867. Accession 25/92. SCT.38 (John Corbett's Testimony); and (Robertson).
11. Ibid., (Corbett).
12. Ibid., (Robertson).
13 Ibid. Accession 25/91. SCT.37 (George Curless's Testimony).
14. Ibid.
15. *Queensland Times*, 28 October 1871.
16. *Towns* v. *J. and J.E. Campbell*, 1867. Accession 25/92. SCT.38 (Corbett).
17. Ibid. Accession 25/91. SCT.37 (Curless).
18. *Statistics of Queensland*, 1865.
19. *Queensland Times*, 28 October 1871.
20. *Votes and Proceedings of the Legislative Assembly of Queensland*, 1871, pp.675-837; and *Queensland Times*, 25 June 1870.
21. *Votes and Proceedings of the Legislative Assembly of Queensland*, 1871, pp.675-837.
22. *Queensland Times*, 28 October 1871.
23. *Votes and Proceedings of the Legislative Assembly of Queensland*, 1871, pp.675-837.
24. These facilities had not saved the mine from being inundated by the flooding of the Brisbane River in 1870, *Queensland Times*, 29 March 1870.
25. Ibid., 28 October 1871.
26. *Queenslander*, 22 April 1871.
27. *Votes and Proceedings of the Legislative Assembly of Queensland*, 1871, pp.675-837.
28. *Statistics of Queensland*, 1872, 1873, 1875 and 1876.
29. *Queenslander*, 11 November 1876.
30. *Australian Dictionary of Biography*, 1851-1890, Vol.6 (Carlton: Melbourne University Press, 1976), pp.294-96.
31. *Sydney Morning Herald*, 17 January 1873.
32. *Queenslander*, 11 November 1876.
33. *Statistics of Queensland*, 1877.
34. *Votes and Proceedings of the Legislative Assembly of Queensland*, 1878, Vol.2, pp.319-402.
35. *Queenslander*, 11 November 1876.
36. A.K. Denmead, "The West Moreton (Ipswich) Coalfield", *Queensland Geological Survey*, Publication 279 (Brisbane: Government Printer, 1955).
37. *Queensland Times*, 28 October 1871.

The Return of Mining to Ipswich

9

The closure in 1862 of Walter Gray's mine[1] left Ipswich with no operation which was being worked on a regular basis. The position changed dramatically as the decade advanced, activity rising quickly from 1865 onwards as Redbank Mine declined in importance. The signs of a renewal of interest in coal mining in the district first manifested themselves some miles to the north east of Ipswich where Sandy Creek follows a sinuous course through the parish of Chuwar before joining the Bremer River (figure 6). Here at one place coal was exposed in the steep north bank of the creek, the outcropping seam later being identified as the Tivoli. In 1865 William Davies[2] purchased Portion 135 Chuwar[3] which included the outcrop and opened a small mine by driving a tunnel into it. By 1868 he was well established and was fulfilling a contract to supply small quantities of coal to the Southern and Western Railway for use as a smithy fuel in their Ipswich workshops.[4] However, cartage was a problem because the town of Ipswich was some five kilometres away and could only be reached by an indifferent road. Moreover, there was no possibility of gaining direct access to the Bremer River which was almost two kilometres distant.[5]

The considerable expense involved in hauling coal overland from a mine to the market led the two Cuff brothers to seek a source of supply nearer to Ipswich than Sandy Creek. Success rewarded their efforts in early 1866 when they sank a shaft just off Coal Pit Gully at a point about half-way between the derelict site of Walter Gray's mine and the recently-completed railway bridge across the Bremer River.[6] The place was probably on, or very close to, the present Portion 16 Ipswich on which North Ipswich school now stands and where a thin band of coaly shale can be identified on the 1899 geological map.[7] The first trial shaft was sunk to a depth of 75 feet (23 metres) where it intersected a seam 7 feet (2.1 metres) thick from which the brothers raised 150 tonnes of coal in a simple iron bucket by means of a windlass.[8] The Cuffs then sank a bigger shaft having a section measuring 10 feet by 5 feet (3 metres by 1.5 metres), and constructed on the surface a single-horse whim 10 feet (3 metres) in diameter for raising the coal.[9] A tramway to the river was also planned but the quality of the seam proved to be a great disappointment, the coal being interbanded with pipeclay and slate. The mine was subject to flooding, much higher quality coal was available fairly close at hand and the operation was therefore abandoned before the end of the year.[10]

The success of the Davies mine on Sandy Creek led to a search for other suitable prospects in that area and in May 1866 an old Scottish miner, Allen Falds, sank a couple of trial shafts immediately across the creek from Davies's site and intersected coal just below the surface.[11] At the same time a local entrepreneurial group led by the Ipswich auctioneer, Mr Hindmarsh, located coal a little lower down the creek and began to work it under the name of the Chuwar Coal Company.[12]

Some years later when the geology of the area became better understood it was realized that the presence of what is now known as the Bishop Fault (figure 6) caused a marked deterioration in the quality of the Ipswich seams on the west bank of Sandy Creek[13] and it is therefore no surprise to learn that Allen Falds's mine and the Chuwar Coal Company ceased operations before the end of 1866.[14] Portion 64 Chuwar, which belonged to George Hutchins, bridged the Bishop Fault and he was able to operate a small pit[15] to the west of the fault,[16] receiving a two year coal contract from the Southern and Western Railway in 1872.[17]

Old Tivoli Pit

At Separation a stretch of unsettled bushland lay to the north and east of the coal allotments on which Walter Gray's mine stood and extended as far as Sandy Creek (figure 6). In 1861 the government divided the area into twenty-two farms, each some 16 hectares in extent, and put them up for sale.[18] In the description of the land which accompanied the plan there is no mention of its potential for coal mining although it had been known for many years that outcrops occurred at various places between the Bremer and Brisbane rivers.[19] The area was described as follows in the plan for the division:

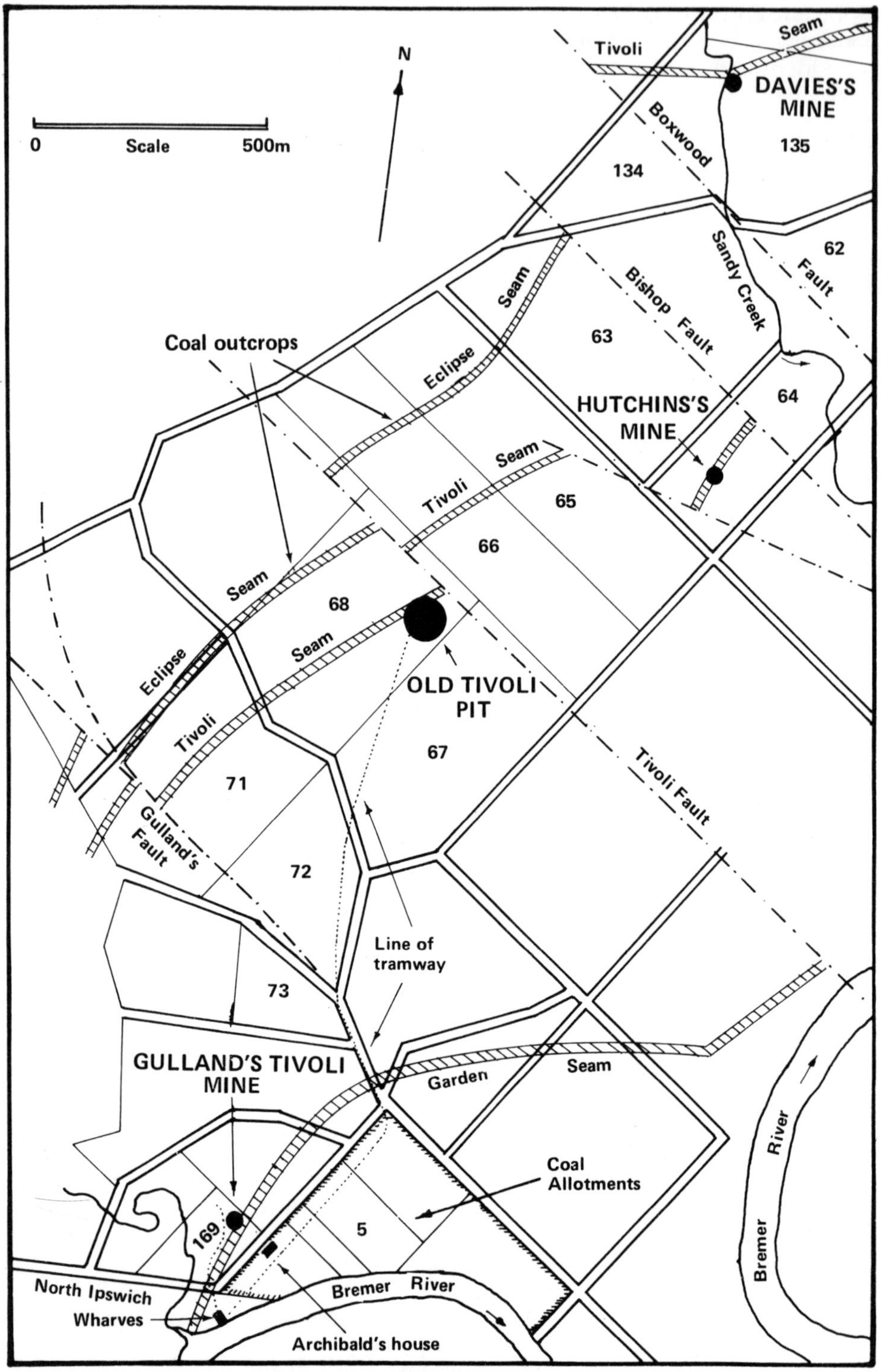

Figure 6. The Tivoli area with the road plan as it existed after subdivision in 1861. Geological features are taken from Cameron's 1899 maps (chap. 9, n. 7). Seams and land Portions not mentioned in the text are omitted for clarity. See Frontispiece for location of the map.

> Parts of Portions 62, 63, 64, 65 and 66 [i.e. those near to Sandy Creek] are suitable for agriculture, the remaining Portions are generally on inferior soil, thickly wooded with small bush, but not heavily with large timber, they are mostly suited for paddocks; the water in the holes shewn on the Plan is apparently permanent. Some of the small Portions near the Town contain good building sites, these Portions are distant from Ipswich from about 1½ to 3 miles, with a good road into it.[20]

The farms were sold off at the upset price applying at that time for agricultural land of £1 per acre ($5 per hectare) and by 1863 most of them had private owners, Portion 68 being purchased by an Ipswich soda-water manufacturer, John Robinson.[21] Whether he was aware that his portion stood astride the outcrop of a high quality seam is not known, but it would be surprising if he had been completely ignorant of the presence of coal. Three years later, in 1866, and impressed by the discovery of coal in a number of places close at hand, he formed a partnership with another local business man, Harry Hooper (plate 25), who had moved to Ipswich from Britain with his brother some years before,[22] and had established a wholesale merchants business trading under the name of Ginn and Hooper. The partners financed the sinking of a trial shaft near the south east boundary of his portion, and the results were sufficiently encouraging for them to sink a second, larger shaft and commence production. By 1867 when it was visited by a reporter of the *Queensland Times* Hooper and Robinson's mine was in full operation.[23]

> Among the foremost of our local industries is the coal mine of Messrs. Hooper and Robinson, North Ipswich, now in full working order...
>
> The scene of operations is situated on the side of a slight incline running down to a blind creek, and is about two miles from town, and nearly one from the part of the river at which the proprietors have erected shoots. When we first visited the place, a few months ago, coal was being raised from the trial shaft at a depth of about 55 feet and the main shaft was being sunk... Over the trial shaft... is constructed a chimney about 20 feet high, for ventilating the mine. A furnace is here [i.e. at its foot] kept burning for the purpose of assisting the ventilation.
>
> Passing a number of improvements in the shape of new huts and fencing, which tend to give a more settled aspect to the place, we come to the main shaft by which, on a previous visit, we had descended into the mine, and through which the coal used to be raised by a horse whim perpendicularly to the surface. The shaft, 10½ft. by 5ft. is now about 60ft. deep. This shaft has been entirely abandoned as an outlet for the coal, and is only used for ventilation, a new mouth to the pit having been opened lower down the hill. Through this opening access to the workings is obtained by oblique instead of by vertical descent. The drive is carried in with an actual slope of one in six, but, as the hill outside slopes upwards at about the

Plate 25. Harry Hooper.

same angle, the declination of the drive from the surface line is really about one in three. The size of the entrance is six feet by six, and these dimensions are preserved for the full length of the main drive—about 200 yards. A train is laid on the floor of the drive,[24] on which run large trollies which receive the smaller trucks as they are filled by the miners in the branch drives or chambers. These trollies are worked by an engine at the mouth of the pit.

This engine is a twelve-horse, with a sixteen-horse boiler, and was formerly used by Mr. Smith at the Town Marie Saw-mills. The engine, when purchased by Messrs. Hooper and Robinson, was a single action, and much expense and loss of time would have been necessary in order to render it reversible, had not the railway authorities kindly permitted the alteration to be effected at their workshops,—the only place at which it would have been done in the colony—a favour for which the proprietors of the mine are not a little grateful.

The engine is a fine horizontal one, by Russell of Sydney,

and is found to work admirably. It is substantially mounted on a strong hard wood frame, and the boiler is set in massive brickwork. A very powerful break [sic] regulates the fly-wheel of the engine. The firing used is the slack from the mine, and this is found superior to piece-coal for raising steam quickly. It cakes well, leaves little ash, and a ton lasts forty-eight hours.

The engine turns a large drum, round which works a wire rope to which the trollies are attached. The drum is 6½ft. in diameter, and makes ordinarily about thirty revolutions a minute, although it can be driven much faster if necessary. The rope is 250 yards long and three-quarters of an inch thick. There is a long shed at the pit's mouth, and a contract has been entered into for covering in the whole of the outside works.

The trollies, on coming out of the drive, are brought up by the engine alongside a stage or landing, at the side of which are the shoots for loading the drays, of which about half a dozen are kept in constant work. The main landing stage is 30 feet by 8, and there is another 16 by 10. The tops of the trollies being exactly on a level with the stages, the small trucks are wheeled from the trollies to the shoots (of which there are three), emptied, and returned to the trollies, by which they are taken back into the mine to be refilled. The quantity of coal that can be thus raised per day of nine hours is 100 tons, but about eighty is the present working. . .

The descent by trolly [sic], and with the assistance of steam power to regulate the speed, is much more pleasant to a visitor than the old plan of going down perpendicularly in a box. In a few seconds one gets enveloped in perfect darkness, and the temperature falls to about 60. At a distance of 115 yards from the mouth, the first branch drive strikes off to the right. This drive extends 300 yards, and abutting on it are a number of chambers, each about five yards square. Tramways for the trucks, which feed the trollies, are laid in all the drives. 35 yards lower down is another drive, and, at a total distance of 200 yards from the entrance the present bottom is reached. This bottom is between 50 and 60 yards perpendicularly from the surface of the ground outside. The roof is a very close-grained freestone. The thickness of the seam is so far uniform at four feet,[25] but the quality of the coal is visibly improving. It does not harden, however, as was expected, but continues rather soft, although very gaseous and rich, and first-rate for steam purposes or for conversion into coke. Some little difficulty is caused by the influx of water in the lower parts of the mine, but it is well kept under. A telegraph cord stretches from the mouth along the whole length of the drives, and by its means constant communication is kept up in a very simple but effective manner.

Where the miners are at work the darkness is rendered just visible by the faint fitful rays of the queer lamps, of which each man carries one on the peak of his cap, and which, one cannot help thinking, must often place the bearer in the somewhat uncomfortable predicament that Aaron found himself in when the oil ran down on his beard, even to the skirts of his garment. The quaint appearance of these grimy half-naked labourers is, however, rather suggestive of a different and, by

> repute, a warmer place than that in which the Jewish high-priest used to officiate. . .
>
> The shoots on the river bank, at which the vessels take in the coal, are distant from the mine about a mile, and the road is in a shocking state, testifying to the heavy traffic to which it is constantly submitted. . . They [i.e. the shoots] are entirely of hardwood, and are capable of holding sixty tons. The drays which convey the coal empty themselves into a large receptacle, from which waggons holding upwards of a ton are filled, and run along a wooden tramway to the shoots on the water's edge, which load the vessels. These are 120 yards from the place where the drays stop, and they overhang the river by several feet. Vessels are loaded with great rapidity from the shoots, and upwards of fifty-five tons is now daily deposited by the drays. We saw a punt receive forty-five tons in about ten minutes. Two men attend to the shoots.
>
> This part of the river is already beginning to assume the appearance of a shipping port on a small scale: a few days since there were four punts and three other vessels alongside waiting for loading. Messrs. Hooper and Robinson now give employment to miners, draymen, puntmen, and labourers—in all nearly seventy men, and about £150 per week is distributed amongst them for wages.

A diagram illustrating the development of Old Tivoli Pit was prepared by James Robertson in 1868[26] (document 10).

The practice of running loaded tubs onto a trolley for haulage to the surface was common in the north of England for a short period in the early years of the nineteenth century but it was soon discontinued because of the complication incurred.[27] A similar change was probably introduced at Hooper and Robinson's mine after 1877, while the drainage problem referred to by the reporter was overcome shortly after the visit by installing a bailing trolley which was later replaced by a steam-driven water pump.[28] Meanwhile the Australasian Steam Navigation Company had tested the coal on their steamers and were so pleased with the results that they contracted for the provision of regular supplies.[29] Local markets for the coal were also secured and in 1867 the future of the mine appeared sufficiently assured for Harry Hooper to purchase Portion 67 Chuwar (which lay immediately down dip from the workings), in order to ensure that the mine had access to adequate reserves.[30] An exploration programme to locate deeper seams was also commenced and in May 1868[31] more coal was intersected at a depth of 120 feet (36 metres) below the worked seam. At the time of the discovery it was referred to as the No.2 seam in order to differentiate it from the upper (or No.1) seam[32] but later it was given various names depending upon the mine which was working it including the Big, Eclipse or Bishop seam.

Quite soon after starting their operations in 1866 Hooper and Robinson must have christened their mine portion of the field where the Coal has already been exhausted the dip and rise is so great that the Miner would have such a difficulty in pushing his Tub or Hutch of Coal up the hill that I cannot believe it would be worked towards the dip. The rooms or Stalls which I observed broken off the levels immediately above the pumping shaft had evidently been worked up the hill and the defendant J. E. Campbell informed me in the presence of the said George Curless that such has been the usual practice in the upper portion of the workings and which I believe to be the fact

15. As to the System and adoption of Adit or day-levels in the first opening up and exploring a Coal Seam I say that at the Burrum Coal field in the Wide Bay district I have been informed by one William Morris the party who began the Workings of the said Burrum Coal field that they adopted this System there, and that all the Coal was exhausted by Horse-tramway. And I also say that Messieurs Hooper and Robinson of the Tivoli Coal Mine near Ipswich after having sunk two Shafts as Marked A and B on the following sketch to their seam of Coal fully fifty feet deep exhausted the Coal to the rise and rather than sink at D a reasonable distance to the Coal they preferred opening up an Adit or day-level as from C where

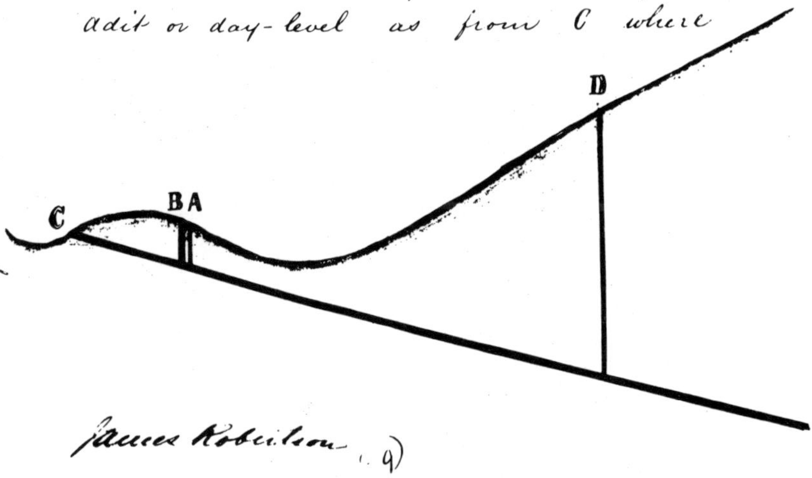

James Robertson (9)

This is the ninth Sheet of the Affidavit of James Robertson sworn this fourteenth day of January in the year of our Lord One thousand eight hundred and sixty eight before me

Reginald Hale
Ipswich

A Commissioner for Affidavits

Document 10. Extract from James Robertson's testimony in 1868 (chap. 9, n. 26) illustrating the development plan at Old Tivoli Pit. The tunnel entrance at "C" was preferred to a new shaft at "D".

the Tivoli and it was referred to by this name in late 1866.[33] The reason for this particular choice is unknown but it must have held a special significance for Harry Hooper because he christened his son with the same name.[34] The No.1 seam subsequently came to be called the Tivoli Seam, and the name spread to the whole of the south western part of the parish of Chuwar and the creek which runs through it. A number of pits which opened later in the area included "Tivoli" in their name and in order to avoid confusion it is convenient to refer to the original Hooper and Robinson mine as the Old Tivoli Pit, the name which was used in the report of the 1881 Select Committee on the tramway which was later constructed to join the mine to the Bremer River.[35]

We now know that the Tivoli Seam is one of a parallel series which outcrops on a north-east/south-west axis about half a kilometre north west of the Tivoli coal allotments (figure 6). It dips on average at about 14° to the south-south-east[36] and contains up to 2 metres of high quality coal. An ash assay as low as 7 per cent makes it greatly superior to practically all other West Moreton coals for steaming purposes, although it has the disadvantage of a high friability. Additionally it produces a good coke although there was no local call for such a product in 1866. Three years elapsed before the demand appeared. In 1869 the Southern and Western Railway decided to modify the fireboxes of its locomotives to burn coke instead of billet wood.[37] When this decision was taken Hooper and Robinson immediately made a bid to supply the coke and the railway awarded them a two-year contract for 50 tonnes per week. The Tivoli Coal and Coke Company, as the partners called their operation at that time, proceeded to construct a battery of ovens in order to fulfil the contract, probably making the bricks from the band of fireclay which had been intersected in the exploration shaft to the No.2 seam.[38] They were described as follows by a visiting reporter in 1870:

> They number twenty-three, and, being built back to back, they form an oblong about seventy yards by ten, and the average height is about seven feet. They are composed of stone and brick, the walls being about three feet thick, and the whole structure has a very substantial appearance. Each oven is circular and dome-shaped, the base being ten feet in diameter . . . coke is made from slack or clean and finely screened coal, and the finer the slack the better the coke . . . Each oven is filled with about five tons, which covers the floor to a thickness of about two feet six inches. It is set on fire very easily by some light wood being placed on top of the slack at the mouth of the oven, and ignited. The only ventilation is by means of small apertures at the mouth of the oven, and the chimney, which is simply a circular opening, about a foot in diameter, at the top of the oven . . . After burning for about seventy hours it is cooled down by means of water . . . The coke is then taken from the oven, and left

exposed to the air for about three hours, so that it may become thoroughly dry, bright, and hard. Five tons of slack produce three tons of coke.[39]

After operating on coke for only six months the railway authorities discovered that the performance of their locomotives was severely impaired and that there was an increased risk of boiler deterioration through corrosion. They therefore commissioned a series of tests from which coal emerged as the most satisfactory fuel. The contract with the Tivoli Coal and Coke Company was cancelled by the government, compensation of £900 ($1800) was paid[40] and in 1872 the company won a three-year contract for the supply of coal instead of coke.[41] Fortunately a market for the coke was found in Adelaide and Melbourne[42] so that the ovens continued in operation and a small export industry was established.

By 1870 the main drive at Old Tivoli Pit was 350 yards (315 metres) in length and the mine was capable of producing 65 tonnes per day.[43] Over 7,000 tonnes of coal were mined in that year, surpassing Goodna's output and making the mine Queensland's largest producer.[44] Robert Archibald, a Scottish miner who had emigrated to Queensland in 1865,[45] was appointed mine manager and at the end of 1871 he bought the site of Walter Gray's old mine overlooking the coal shoots which the company had installed at the mouth of Tivoli Creek.[46] Here he built a comfortable house (plate 26), perfectly placed for oversee-

Plate 26. Robert Archibald's house, Tantivy Street, Ipswich. (Photographed in December 1979.) See also fig. 6, p. 108 and plate 33, p. 126.

ing coal loading operations on the Bremer River. In addition Messrs Hooper and Robinson felt that a younger man was needed in the business to assist with commercial matters and their choice fell on James Gulland.

James Gulland, a Scotsman by birth, had emigrated to Queensland in 1864 at the age of twenty-eight.[47] Here he obtained a position in the public service as a clerk,[48] rising over the following four years to secretary to the postmaster general's department.[49] In July 1870 he relinquished this post to become a partner with Hooper and Robinson in their mine,[50] quickly becoming the driving force in the Tivoli Coal and Coke Company. By 1872 he had assumed full control,[51] allowing Hooper and Robinson to sell out their shares of the mine to him[52] and move to Gatton,[53] where they developed separate interests in the pastoral industry.[54]

Gulland's first task was to secure the future of the mine. The No.2 seam which had been located in 1868 was a considerable disappointment, being of a poorer quality than the No.1 (or Tivoli) seam on which the mine relied for production.[55] The output from the No.2 seam had to be curtailed and Gulland cast around urgently for better alternatives, purchasing the adjacent Portion 72 Chuwar for its coal reserves in 1872[56] after another seam below No.2 had been discovered and discarded. Later a new tunnel entrance was opened into the Tivoli Seam close to Francis Road[57] but Gulland also resolved to start a second mine in the district.

At the time of the closure of Walter Gray's mine in 1862 it was well known that coal outcropped near at hand on the east bank of Tivoli Creek.[58] Three of the twenty-two small farms which had been offered for sale by the government in 1861 and which lay immediately to the north of Walter Gray's mine bordered on Tivoli Creek and must have shown surface indications of the presence of coal. However they were not taken up and in 1870 a redivision took place.[59] The newly designated Portion 169 Chuwar lay immediately to the north of the site of Walter Gray's mine (figure 6) and a certain Timothy Hegerty purchased it, selling it immediately to John Archibald who was the brother of Robert, the manager of the Old Tivoli Pit.[60]

The brothers carried out a careful exploration of the portion, coming to the conclusion that it was well endowed with coal[61] although they do not appear to have exploited their discovery. Late in 1873 John Archibald left the district to try his luck elsewhere,[62] but before doing so he sold Portion 169 to his brother Robert who then leased it to James Gulland.[63] A tunnel mine was immediately opened, a short tramway was built to the company's shoots[64] at the mouth of Tivoli Creek and the operation became known as Gulland's Tivoli Mine.[65] When the railway contract expired at the end of 1874 Gulland offered coal from both mines in open competition with samples from other producers. The railway conducted a series of tests on them, finally choosing the sample from Gulland's Tivoli Mine and subsequently signing a three-year contract for the supply of 110 tonnes per week. An expenditure in the following year by the Railway Department of £56.11s.7d. ($113) for the "Tramway, Gulland's Coal Pit", may have been for building a spur on the tramway where the coal could be held until required at the Ipswich railway yards. When the geological surveyor for Queensland, Augustus Gregory, visited the Ipswich coalfield in 1876 the high quality of the coal from Gulland's Tivoli Mine led him to infer that it was working the Tivoli Seam, but as the geology of the area became better understood it was realized that the workings were in fact located on the Garden Seam which was parallel to, but lay above, the Tivoli Seam.[66]

Meanwhile the problem of carting coal from Old Tivoli Pit to the river along indifferent roads was adding about 1s.6d. per ton (15 cents per tonne) to the price of the coal,[67] and this extra cost caused Gulland considerable concern. The need for a tramway had been recognized as early as 1866[68] but there were problems in selecting a suitable route. An unfortunate feature of the terrain was that a ridge barred the way from the mine to the river. Moreover, Gulland did not own a corridor of land along which the tramway could be laid, but at the end of 1873 he purchased leases which overcame this obstacle[69] except in the middle section where he arranged to lay the track along the side of the public road. The wooden rails on the new tramway were fitted with flanges[70] and were set at a spacing of 15 inches (375 millimetres).[71] The new system was put to work in 1876 and was described in some detail shortly afterwards by a visiting reporter.

> The Tivoli railway is designed to carry coal, which until recently was carried in drays. The total length of the work is about one mile and a quarter. Starting from the coal shoots at the bank of the river, the line at once faces a very steep ridge for about 600 yards, the whole length of which is a series of cuttings and embankments. On this a double tramway track of timber has been laid, and a continuous string of cars, each carrying 7 cwt. of coal, moves along. The down train being loaded, and the up train being light, and both being attached to an endless wire rope, the position has been taken advantage of to obtain all the necessary motive power from the cars on the down grade.
>
> At the top of the double line of rails the rope, which is of wire, passes round a gutted wheel. This wheel is horizontal, and is fixed upon a strong framework of hardwood posts. A long lever is made to act as a break [sic] upon the wheel, and by this means one man controls the whole up and down traffic of the 600 yards passing over the side of the steep hill.
>
> The traction rope is carried on the trucks, and by an

ingenious contrivance it clips into a slotted piece of iron at each end of each truck. The system of working is to hook on each loaded truck as it comes to the top of the line from the pit, allowing each to go forward as it is attached to the rope, and at the same time taking off an empty truck from the up line. As the loaded trucks reach the bottom of the incline, they are, one by one, disengaged by simply lifting the rope out of the slots. Then the coal is discharged into the shoots, or into a vessel, if there is one on the berth, and the truck is hooked on to the train going upwards. Three men attend to this portion of the work, without any horse, steam, or other extra power.

The Tivoli coal pit is down on the other side of the ridge from that facing the river, about 1600 yards, and ingenious as the means employed on the river side are to send down the loaded and get up the empty trucks, the contrivance for getting the loaded trucks up from the pit outlet to the top of the ridge is more ingenious still. As on the other side, a double line of wooden tramway has been laid down, and an endless wire rope, is used for the hauling. The length of tramway at this side is about 1600 yards, 600 of which are worked by wire rope, and the remainder by horses. The grade is steep as 1 foot of rise in 12 of track near the pit; towards the top it is less steep. To raise a continuous string of loaded coal trucks up such an incline, even with the counter-balancing weight of the empty trucks on the down line, requires considerable power. This power is obtained from the engine used for hoisting or rather dragging coal from the pit. From the surface the pit goes down an incline of 1 foot in 5, up which the coal is dragged by a single wire rope. Three loaded trucks are brought up each trip, and taking advantage of the power of the empty trucks going down into the pit, a rope of wire is passed from the drum to which the descending trucks are attached, and from the drum a wheel is set in motion that drags the loaded cars up the incline from the mouth of the pit.

Formerly the descent into the pit was controlled by a powerful brake; now the power absorbed by that break [sic] is transferred to a wheel from which passes the endless wire rope that carries the coal up the side of the steep ridge from the pit, and brings back the empty trucks. The adaption of power is clever and decidedly skilful. Mr. Robt. Archibald, the manager at Tivoli, engineered the work, and it says much for his faculty for using available material.

In all about one mile and a quarter of line is constructed from the mouth of the pit at Tivoli to the bank of the Bremer River. The route, though short, is over difficult country. Timber is used for the rail or tramway. The wheels of the trucks are not flanged, they are plain; they travel upon flat rails of timber, with edgings on the inside to keep the wheels in position and on the track. Wheels and axles are wedged and travel together as in the case of ordinary railway wheels and axles.

The total cost of the work at Tivoli is £1500. The traffic that is carried over it kept seven horse teams and twenty-one horses going, and hard nagging work they must have had, particularly in wet weather, for the road, in addition to being a series of severe pinches, is over material not easily kept in order for heavy traffic.[72]

In 1876 the length of the original underground drive had been extended to 600 yards (540 metres), matching exactly the length of the ropeway running from the portal to the top of the ridge and thus permitting the construction of the ingenious haulage system described by the reporter. Annual coal production climbed quickly, the combined output of the two pits reaching 27,000 tonnes in 1877. Criticism also began to mount at the inconvenience of having a tramway running alongside a public road and crossing it at three places. The antagonism reached a climax in 1881 with the expiry of the lease on the land of a rival mine owner through which the tramway passed. The obdurate owner, John Eastwood, refused a new term of the lease forcing Gulland to appeal to the government who appointed a select committee to investigate the matter. The committee found in favour of Gulland and the tramway was permitted to remain under the terms of special parliamentary legislation.[73]

Gulland continued to enlarge the area of coal reserves available to the Old Tivoli Pit by purchasing Portion 71 Chuwar in 1881 and taking out a thirty year lease on Portion 73 Chuwar in 1883.[74] Notwithstanding these additions, Gulland was astute enough to recognize that a mine which could be easily connected to the growing railway system of the colony would always have an advantage over one which could not. The Ipswich to Brisbane line which had been opened in 1874[75] and which ran along the south side of the Bremer River was inaccessible to the Tivoli pits and he therefore began a search for a suitable site for a coal mine within easy reach of the railway. His final choice fell on portions, 265, 266, 267, and 268 Goodna, containing the old Redhill Mine which had lain abandoned since Campbell and Towns had explored the area in the late 1860s. (See pages 131, 132). From 1879 onwards and through a series of complex land deals made in collaboration with a relative of his wife, Robert Russell Smellie,[76] he shifted his interests south of the Bremer and wound down the operations at Tivoli.[77] When Tenison-Woods visited Ipswich in 1883 the move had been completed and the site of the Old Tivoli Pit was deserted.[78] And so it remained until the turn of the century when the portions were sold to the Stafford brothers and a new phase of mining commenced on the old site.

Other Operations North of the Bremer River

The success of Old Tivoli Pit stimulated a succession of other operations in the district, some of which were successful but many of which were not. A great problem was that the geology of the district was not understood. It was not appreciated, for example, that there was a series of parallel coal seams dipping to the south east which outcropped in the Tivoli area (figure 6) and differed considerably in quality. The existence and location of two or three important faults were not obvious from the surface but they too greatly affected coal quality.

The portions 65 and 66 Chuwar lying immediately to the north east of the Old Tivoli Pit were put up for sale on a number of occasions[79] after their original purchase in 1864.[80] Although they were situated between the mines at Tivoli and Sandy Creek (a point which the contemporary advertisements stressed heavily), no buyer could be found for them until well into the 1870s when they came under the control of Messrs Blond and Wright[81] and produced substantial tonnages of coal for the next thirty years.

The presence of coal on the Waterstown Estate which adjoined Old Tivoli Pit to the south east was established in 1866 but despite attempts by two adventurers, Hay and Schofield, to start operations in 1872,[82] development had to await the successful passage through the parliament in 1874 of the Waterstown Coal Leases Bill. The Bill permitted the land to be leased by the trustees of the Moffatt family for mining purposes, and for a royalty to be charged on all coal raised.[83]

In 1873 John Eastwood purchased Portion 70A Chuwar near Mihi Creek as a mineral selection[84] and the Eastwood Mine was opened shortly afterwards. Thus by 1875 there was the potential for a rapid expansion in the production of coal in the North Ipswich area. However, the Tivoli Coal and Coke Company could satisfy all the current market needs, and the cost of carting coal from any new operation which did not have ready access to the river discouraged competitive development. The establishment of an export market and the construction of a connecting railway line to the Tivoli area which were both essential for a substantial increase in production were still two decades away.

The Sites in 1980

Of the sites already described possibly Davies's mine has changed the least over the years. The entrance is still visible in the rocky bank of Sandy Creek (plate 27), and Denmead reported as recently as 1955 that it was possible to squeeze into the entrance.[85] The other Sandy Creek mines were so short-lived and their positions so inadequately defined in the first place that they have never been located with any degree of certainty. Hutchins's mine on Portion 64 is marked as an old tunnel in Denmead's 1955 survey[86] but the land has subsequently been levelled, subdivided and built on so that no surface feature of it now remains.

Plate 27. Davies's mine on Sandy Creek. (Photographed in June 1980. The scale divisions on the measuring rod are each 100mm.)

Entrances in the vicinity of Old Tivoli Pit are marked on Mines Department survey maps released in 1899, 1923, 1955 and 1957.[87] However agreement on the exact positions of the openings is very poor and a careful survey of the site in 1979 failed to locate any entrances which corresponded exactly with the earlier surveys. An additional confusing factor is that extensive spoil heaps which mark the site of the mine were levelled after the second world war for safety reasons and used for filling in the shaft and tunnel entrances.

Plate 28. View of Old Tivoli Pit across the general mine area from the blacksmith's hearth to the coke ovens. (Photographed in December 1979.)

The author's 1979 survey revealed two openings near the north east boundary of Portion 68. One was a stone-lined depression 1.75 metres square lying on the south flank of the small ridge on which the mine lies. The other was a circular pit situated 90 metres to the north of the first depression. Its position corresponds exactly to that of an old shaft marked on a pre-1912, two chains to the inch, map of the underground workings of the Eclipse seam at Rothwell Haigh. Thus the circular pit is probably the site of the vertical shaft which was sunk to the No.2 seam in 1868. No trace was found of the trial shaft and first ver-

Plate 29. Coke ovens at Old Tivoli Pit. (Photographed in December 1979. The scale divisions on the measuring rod are each 100mm.) See also fig. 14, p. 172.

tical shaft which the reporter described in 1867 and which are shown diagrammatically in a section of the pit drawn by James Robertson to illustrate his testimony to the 1867–68 Towns and Campbell law case (document 10, page 113).

The row of twenty-three coke ovens has decayed into a tree and bush covered barrow, strewn with sandstone blocks and bricks. A survey and excavating programme in 1979 confirmed the accuracy of the press report which described their construction in 1870 (plates 28, 29, and 30).

Artefacts discovered around the smithy hearth in the mine area include pre-1865 coinage, a variety of horseshoes and parts of broken trolley wheels. These are unflanged and one of them has the charred remains of a wooden wedge clinging to the inside of the square-sectioned hub.

Plate 30. Excavated brick floor of coke oven at Old Tivoli Pit. (Photographed in December 1979. The scale divisions on the measuring rod are each 100mm.) See also fig. 14, p. 172.

The line of the tramway incline can be easily followed from the pit area along a small stone-faced embankment near its foot and up a shallow cutting to Francis Street (plates 31 and 32). Here the comparatively level horse-hauled section commenced, and a shallow cutting can be located behind the houses on Portion 72 Chuwar where the tramway passed across Gulland's land and thus avoided the highest point of the ridge. The following section along Tivoli Street to the descending incline is now built up and all signs of its route have disappeared. The cuttings at the top of the incline leading down to the shoots have been filled in but the embankment across

Plate 31. Cutting on line of tramway decline to the Old Tivoli Pit. (Photographed in September 1979.)

Plate 32. Tramway decline to the Old Tivoli Pit showing one of the iron drains which pierced the embankment near Francis Street. (Photographed in December 1979. The scale divisions on the measuring rod are each 100mm.)

Portion 5 Chuwar which belonged to John Eastwood and which led to the select committee of enquiry in 1881, is clearly visible. In the area where the tramway crossed the site of Walter Gray's mine behind Robert Archibald's house and reached the coal shoots adjacent to Tivoli Creek the line can still be followed despite subsequent redevelopment (plate 33).

Plate 33. Old Tivoli Pit tramway decline to the Bremer River. Robert Archibald's house is on the left and the tramway ran along the right hand boundary. (Photographed in December 1979.)

The site of "Gulland's Tivoli Mine" is clearly visible as a south easterly facing depression at the top of Hastie Street. The spoil heap can also be identified together with part of the line of the tramway leading to the shoots at Tivoli Creek (plate 34).

The whole of the site of the Old Tivoli Pit is scheduled for reworking as an open cut mine. A full recording of the surface features left by the original operation has therefore been made and lodged with appropriate bodies for permanent retention.

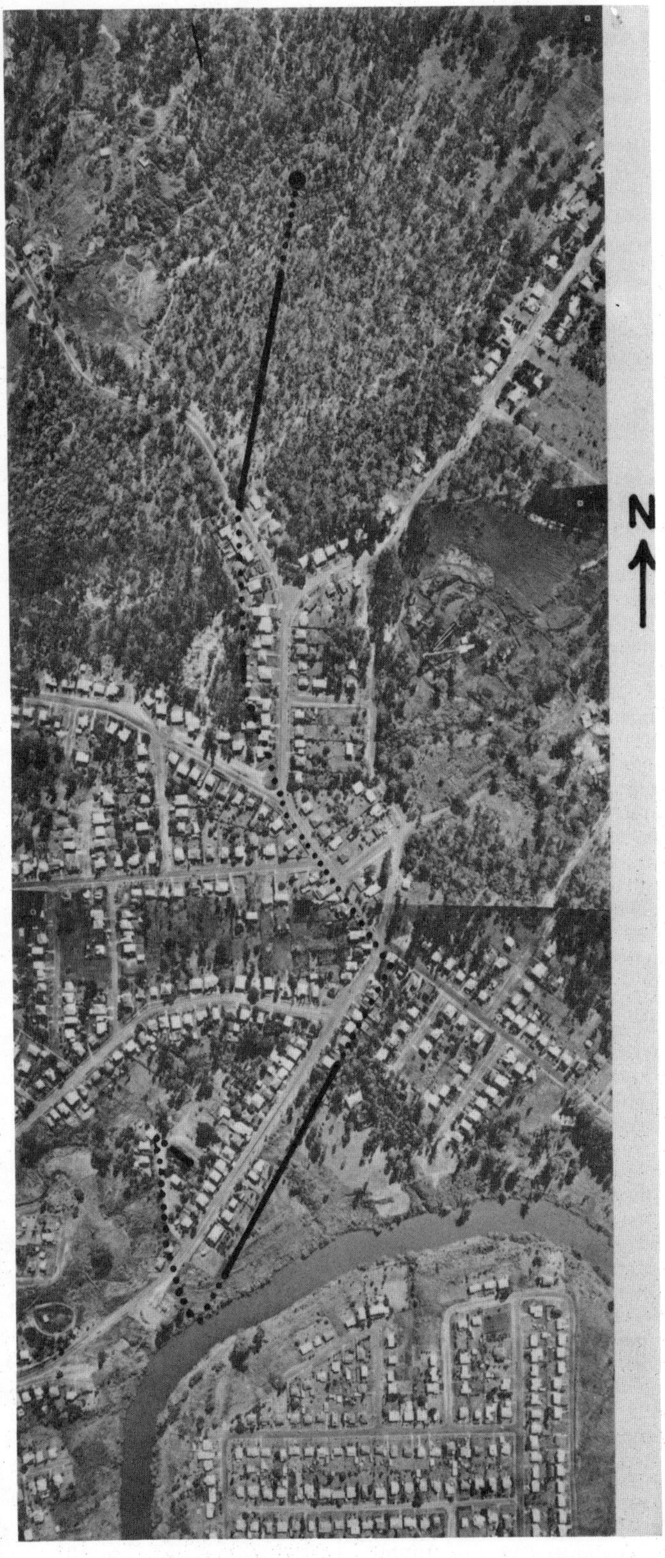

Plate 34. Mosaic of North Ipswich showing the sites of Old Tivoli Pit and Gulland's Tivoli Mine, and lines of the tramways joining them to the coal wharves on the Bremer River at Tivoli Creek. Where the tramways could still be identified in 1980 the lines are solid. (Photographed in May 1979.)

Notes and References for Chapter 9

1. *Statistics of Queensland*, 1862, p.52.
2. In some documents "Davies" is contracted to "Davis", but the references are to the same mine. See notes 5 and 7.
3. Queensland Government Titles Office, Vol.15, Folio 233, 1 April 1865.
4. *Votes and Proceedings of the Legislative Assembly of Queensland*, 1869, Vol.1, p.519.
5. *Queensland Times*, 17 April 1866.
6. Ibid.
7. W.E. Cameron, "Geology of the West Moreton or Ipswich Coalfield", *Queensland Geological Survey*, Publication 147 (Brisbane: Government Printer, 1899).
8. *Queensland Times*, 17 April 1866.
9. Ibid., 19 July 1866.
10. Ibid., 20 September 1866.
11. Ibid., 10 May 1866.
12. Ibid., 12 May 1866.
13. *Votes and Proceedings of the Legislative Assembly of Queensland*, 1900, Vol.2, pp.300-339.
14. *Queensland Times*, 20 December 1866.
15. Ibid., 6 February 1872.
16. Cameron, "Geology of the West Moreton or Ipswich Coalfield".
17. *Queensland Government Gazette*, 1872, Vol.13, p.572.
18. Queensland Government Department of Mapping and Surveying, Brisbane, Map M31.47.
19. Ibid., Map MT56.
20. Ibid., Map M31.47.
21. Official Post Office Directory of Queensland (Brisbane: Government Printer, 1868); and Queensland Government Titles Office, Vol.9, Folio 114, Deed 6776, 9 December 1863.
22. *North Australian*, 6 July 1858; and 21 September 1858.
23. *Queensland Times*, 20 December 1866; and 19 October 1867.
24. The track is given as made of timber 4 inches (100 millimetres) square, with a gauge of 3 feet (2.7 metres) in *Queenslander*, 21 October 1871.
25. R. Daintree, *Queensland, Australia* (London: G.Street, 1873), p.78, gives the thickness of the seam as 3 feet (0.9metres); *Votes and Proceedings of the Legislative Assembly of Queensland*, 1876, Vol.3, pp.387-95, gives it as 5 to 6 feet (1.5 to 1.8 metres) thick; *Queenslander*, 18 June 1870, stated that a 3 foot and a 1 foot seam had combined to give a thickness of 4 feet 2 inches (1.3 metres).
26. *Towns* v. *J. and J.E. Campbell*, 1867. Queensland State Archives. Accession 25/92. SCT.38 (James Robertson's Testimony).
27. F. Atkinson, *The Great Northern Coalfield* (London: University Tutorial Press, 1968), p.30.
28. *Queenslander*, 19 January 1878; 21 October 1871; and 19 January 1878.
29. *Queensland Times*, 20 December 1866.
30. Queensland Government Titles Office, Vol.3, Folio 143, 15 May 1867.
31. *Queensland Times*, 16 May 1868.
32. *Votes and Proceedings of the Legislative Assembly of Queensland*, 1871, pp.675-837; and 1878, Vol.2, pp.319-402.
33. *Queenslander*, 22 December 1866.
34. Information from Mr Fred Hooper.
35. *Votes and Proceedings of the Legislative Assembly of Queensland*, 1881, Vol.2, pp.977-1005.
36. Cameron, "Geology of the West Moreton or Ipswich Coalfield".
37. *Votes and Proceedings of the Legislative Assembly of Queensland*, 1876, Vol.3, pp.387-95; 1871, pp.675-837; and 1871, pp.661-73.
38. Daintree, *Queensland, Australia*, p.78.
39. *Queenslander*, 18 June 1870.

40. *Votes and Proceedings of the Legislative Assembly of Queensland*, 1871, pp.661-73; and 1871-72, p.31.
41. *Queensland Government Gazette*, 1872, Vol.13, p.572.
42. *Votes and Proceedings of the Legislative Assembly of Queensland*, 1871, pp.675-837.
43. *Queenslander*, 18 June 1870.
44. *Votes and Proceedings of the Legislative Assembly of Queensland*, 1871, pp.675-837.
45. Ibid., 1881, Vol.2, p.977.
46. Queensland Government Titles Office (Old System), Book 37, No.143, 3 November 1871.
47. *Queensland Times*, 18 February 1904.
48. *Queensland Blue Book*, 1867.
49. *Queensland Government Gazette*, 1868, Vol.9, p.204.
50. *Queensland Times*, 1 September 1870.
51. Ibid., 21 November 1872.
52. Queensland Government Titles Office, Vol.183, Folio 3, 3 January 1872; Folio 4, 14 February 1872; Folio 5, 30 April 1872; Folio 6, 3 January 1872; Folio 7, 14 February 1872; and Folio 8, 30 April 1872.
53. *Pugh's Almanac* (Brisbane: Thorne and Greenwell, 1875), pp.113-120.
54. Information from Mr Fred Hooper.
55. *Votes and Proceedings of the Legislative Assembly of Queensland*, 1871, pp.675-837.
56. Queensland Government Titles Office, Vol.185, Folios 123 and 125, 26 April 1872.
57. *Votes and Proceedings of the Legislative Assembly of Queensland*, 1878, Vol.2, pp.319-402; referred to as "Gulland's Tivoli Tunnel" by A.K. Denmead, "The West Moreton (Ipswich) Coalfield", *Queensland Geological Survey*, Publication 279, Map No.1 (Brisbane: Government Printer, 1955).
58. *Brisbane Courier*, 24 October 1862.
59. Queensland Government Department of Mapping and Surveying, Map S31.423.
60. Queensland Government Titles Office, Vol.167, Folio 149, 25 November 1870; and Vol.167, Folio 149, Transfer 24089, 9 December 1870.
61. *Queensland Times*, 6 February 1872.
62. D.B. Waterson, *A Biographical Register of the Queensland Parliament, 1860-1929* (Canberra: Australian National University Press, 1972), p.5.
63. Queensland Government Titles Office, Vol.213, Folio 36, 22 December 1873; and Vol.213, Folio 36, Lease 32476, 2 January 1874.
64. Queensland State Archives. Railway Map C.B.3.
65. Cameron, "Geology of the West Moreton or Ipswich Coalfield".
66. *Votes and Proceedings of the Legislative Assembly of Queensland*, 1876, Vol.3, pp.683-706; pp.565-94; and pp.387-95.
67. Ibid., 1871, pp.675-837.
68. *Queenslander*, 22 December 1866.
69. Queensland Government Titles Office, Vol.174, Folio 96, 1 November 1873; Vol. 169, Folio 166, 1 November 1873; Vol.173, Folio 57, 19 November 1873; and Vol.173, Folio 56, 1 November 1873.
70. The flanging is variously stated as being on the inside of the rail (*Queenslander*, 19 January 1878), and on the outside in *Votes and Proceedings of the Legislative Assembly of Queensland*, 1881, Vol.2, pp.977-1005. The builder of the line, Robert Archibald, said that the flange was on the outside, *Votes and proceedings*, 1877, Vol.1, p.397.
71. Ibid., 1881, Vol.2, pp.977-1005.
72. *Queenslander*, 19 January 1878.
73. *Votes and Proceedings of the Legislative Assembly of Queensland*, 1876, Vol.3, pp.387-95; 1878, Vol.2, pp.319-402; 1877, Vol.1, p.397; and 1881, Vol.2, pp.977-1005.

74. Queensland Government Titles Office, Vol.28, Folio 109, 13 May 1881; and Deed of Grant 8499, 21 May 1864, Lease 85348, 29 September 1882.
75. *Votes and Proceedings of the Legislative Assembly of Queensland*, 1876, Vol.3, pp.565–94.
76. Queensland Government Titles Office, Vol.116, Folios 209 and 210, 12 July 1867; and Vol.116, Folios 209, 13 June 1882, and 210, 21 January 1884.
77. W. Fryar, Annual Report to Department of Mines, 1899 (Brisbane: Government Printer), pp.156–64.
78. J.E. Tenison-Woods, *The Coal Resources of Queensland* (Brisbane: James C. Beal, Government Printer, 1883).
79. *Queensland Times*, 3 July 1866; and 11 February 1869.
80. Queensland Government Titles Office, Vol.37, Folio 9, 21 May 1864.
81. *Votes and Proceedings of the Legislative Assembly of Queensland*, 1876, Vol.3, pp.387–95.
82. *Queensland Times*, 19 June 1866; and 21 March 1872.
83. *Votes and Proceedings of the Legislative Assembly of Queensland*, 1874, Vol.2, p.1052.
84. Queensland Government Titles Office, Vol.239, Folio 44, Mineral Selection 2541, 29 November 1873.
85. Denmead, "The West Moreton (Ipswich) Coalfield", p.54.
86. Ibid.
87. Cameron, "Geology of the West Moreton or Ipswich Coalfield", Publication 147; W.E. Cameron, "Geological Map of Ipswich and Bundamba Coalfields", *Queensland Geological Survey*, Publication 271, New Series; Vol.8, Sheet 2 (Brisbane: Government Printer, 1923); Denmead, "The West Moreton (Ipswich) Coalfield", Publication 279, Map No.1; and *Queensland Geological Survey*, "Coal Resources, West Moreton (Ipswich) Coalfield", Publication 286, Part 5, Figs 1, 2 (Brisbane: Government Printer, 1957).

Operations South of the Bremer River

10

From the earliest days of settlement the main road linking Brisbane to Ipswich and the hinterland ran immediately to the south of the Brisbane and Bremer rivers. In 1860 shortly after Separation the government subdivided some 640 hectares of second-class pastoral land[1] which lay astride this road immediately to the east of the boundary of the new municipality of Ipswich and to the south of the Bremer River.[2] Although sales of the forty-two portions were slow, particularly where the land was at a distance from both the river and the road, it was prospected by some of the hopeful parties of immigrants who were pouring into the new colony of Queensland.

Among the prospectors were two miners and in 1866 they quietly sank a shaft in a creek bed which ran from east to west near the boundary between portions 265 and 268 Goodna (figure 7). At a depth of 40 feet (12 metres) they intersected a seam of coal over 6 feet (1.8 metres) in thickness which appeared to be of excellent quality. Instead of protecting their discovery by applying for a coal mining lease under the 1865 regulations they foolishly talked freely about their find. The news came to the ears of the Campbells at Redbank who immediately purchased the portions for the Campbell and Towns partnership and ejected the unfortunate miners.[3]

A second shaft which the partnership proceeded to sink near to the original opening intersected some 24 feet (7 metres) of coal, although it was broken by stone bands varying from ½ inch to 6 inches (12 to 150 millimetres) in thickness. Moreover the seam appeared to dip steeply to the east and this was confirmed by boring a hole 150 yards (135 metres) to the east of the new shaft where it

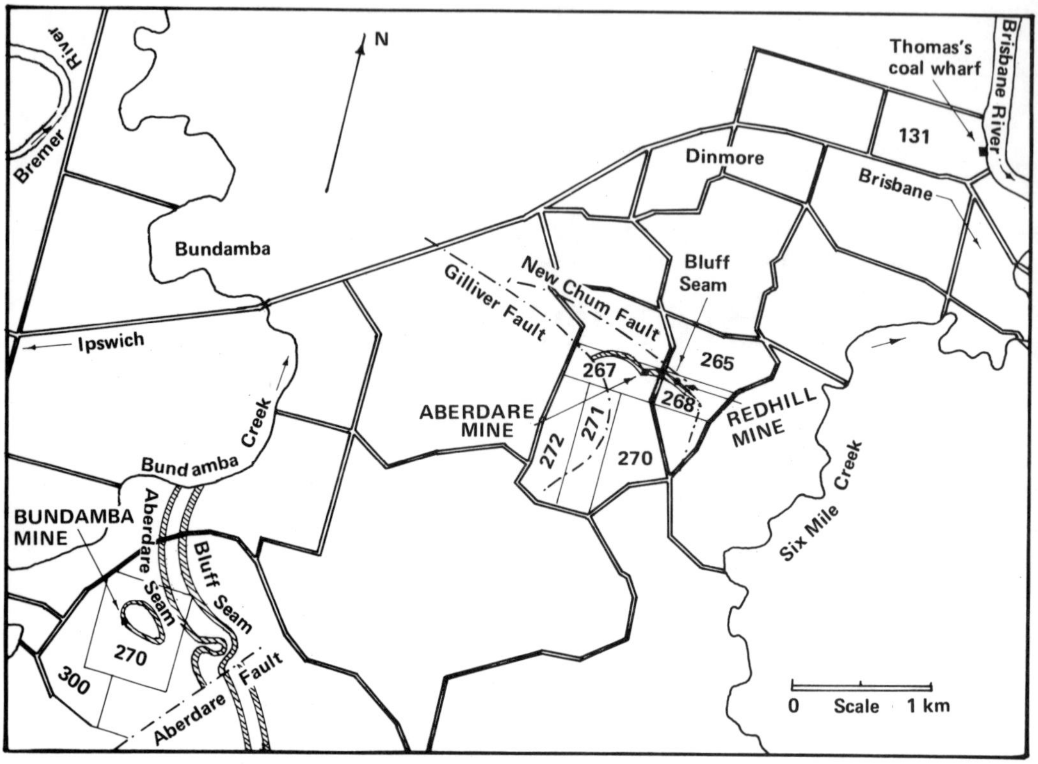

Figure 7. The Bundamba and Dinmore area, with the road plan as it existed in 1870. Geological features are taken from Cameron's 1899 maps (chap. 10, n. 39). Seams and land portions not mentioned in the text are omitted for clarity. See Frontispiece for location of the map.

intersected the coal at a depth of 164 feet (49 metres). After naming the colliery Red Hill and taking a section 5½ feet (1.65 metres) thick out of the coal for a distance of about 20 yards (18 metres) from the bottom of the shaft to the dip the mine was abandoned. The quality was claimed to be very similar to that at Redbank but the long haulage distance to the river and the steepness of the seam made further development unattractive.[4] Nine years later in 1876 when the government geological surveyor, Augustus Gregory, visited the area the abandoned workings were still visible.[5]

In the intervening period one of the district's most persistent and successful coal miners, Lewis Thomas, had realized the significance of the position of the Red Hill Mine and had opened the Aberdare Mine on an adjacent portion lying to the west of the original workings. Before following its fortunes, however, it is necessary to trace the vicissitudes in the career of this remarkable man which led to his becoming the proprietor of the Aberdare Mine in 1870.

Lewis Thomas

Lewis Thomas, who was born in 1832 at Talybont in Cardiganshire, Wales, migrated to Australia at the age of

Plate 35. Lewis Thomas. (Courtesy John Oxley Library.)

twenty-six (plate 35). He had left school at the age of nine to work in a woollen mill and subsequently completed his education by taking a wide variety of jobs, including working as a miner in the coal and iron ore mines of South Wales.[6] "Of a rugged personality, possessed of unflagging energy and an indomitable spirit which enabled him to persevere in the face of apparently insurmountable obstacles, he was just the stamp of man required by this distant part of the great British Empire in the strenuous days of its youth".[7] After failing on the Victorian gold diggings he moved to Queensland where in 1861 he began working at Redbank as a skilled miner.[8]

By 1864 Thomas had recognized that his chances of advancement at Redbank under the Campbells were limited and he looked elsewhere for a suitable opportunity to further his career. He found it in Queensland's first railway line which was being built between Ipswich and

Toowoomba. One of the major obstacles in the path of the line was the Little Liverpool Range and the plans called for a tunnel to breach it. Thomas with considerable courage and self confidence made a bid for driving the western half of the Victoria tunnel which was accepted by the main contractors.[9] By early 1865 he had completed the sub-contract although not without experiencing various difficulties including roof falls which claimed the life of one of his workmen (plate 36).[10]

Plate 36. Western portal of the Victoria railway tunnel in the Little Liverpool Ranges. (Photographed in July 1980.)

In the mid 1860s the minor success of Davies's mine on Sandy Creek[11] stimulated coal prospecting activity to the north and east of Ipswich. Thomas forsook tunnel construction to join the search and began looking for a suitable site in the Tivoli area. A prospect was found near Waterstown[12] but it soon failed and by early 1866 he was once more looking for an outlet for his talents. He found it in a short-lived partnership with John Malbon Thompson.

The contrast between the early careers of Thomas and Thompson could hardly have been greater. Whereas Thomas was a self-taught, impecunious migrant, Thompson came from a well-connected Sydney family.[13] In 1855 he had been admitted to the Supreme Court of New South Wales as an attorney and solicitor,[14] moving to Ipswich two years later where he bought the practice of a deceased local solicitor, James Walsh.[15] Thompson's interest in coal mining probably dated from 1861 when he successfully defended the Redbank miners against John Campbell in a Masters and Servants Act case which resulted from a strike at the mine (see Chapter 5).[16] His peripheral interest was transformed into a practical reality in August 1865 when he took out a coal mining lease on 100 acres (40 hectares) of land at Bundamba[17] which was being sold by the government as Portion 270 Goodna (see figure 7 on page 132).[18] But Thompson's talents fitted him for politics rather than for industry, business or commerce. His qualities of competence combined with a courteous and punctilious manner[19] appealed to the Ipswich electors who returned him as their member of the Legislative Assembly in 1868. Subsequently he became Secretary of Lands and Minister for Public Works when his abilities as a conciliator and mediator earned him the reputation of being "a liberal whose moderation has been interpreted as weakness".[20]

Under the 1865 regulations[21] the terms of the mining lease on Portion 270 Goodna required Thompson to commence operations within a period of twelve months from the issue of the grant and in 1866 he therefore cast around for an experienced, practical miner whom he could employ on the work. A suitable person was not easy to find, and instead he agreed to take Lewis Thomas into partnership. Thomas set to work with great speed and energy to open a mine on the portion and by the middle of 1866 the partners were advertising the sale of high quality coal from their operation at Bundamba.[22]

Bundamba Mine

Portion 270 Goodna stands on the edge of the hills which rise gently to the south from the Bremer River some three miles to the east of the centre of Ipswich. A steep-sided valley runs across the portion commencing behind Blackstone Hill and leading in a westerly direction to Bundamba Creek. By a fortunate quirk of nature the valley is just deep enough along practically its whole length to intersect an almost horizontal high-quality coal seam (later termed the Aberdare), so that in 1866 a mine could be started almost anywhere by driving directly into the sides of the valley at the appropriate level. Six months after the mine had commenced production the operation was described by a visiting reporter.

> Next to Tivoli coal mine in extensiveness of operations, and of equal importance as regards the quality of the coal and the workable area, is the Bandambah mine, at present worked under the joint proprietary of Messrs. Thompson and Thomas. This scene of industry is situated a little more than two miles from town, in the midst of a number of precipitous gorges. The road leading to it is a very indifferent one, and in wet weather almost impassable. Leaving the Government road a short distance beyond the residence of Mr. J.M. Thompson, a road strikes off to the right, out around the edge of the hills. This is about three-quarters of a mile in length, and leads to the entrance to the mine, at a bight where two hills meet. There is no sinking, as the seam was struck at the surface, and the whole work consists in driving. The main drive is about six feet in width, with a height of four feet, and its present extent is about 80 yards from the mouth. At about 25 yards is a heading which extends about 60 yards, off which are several smaller headings, which at the length of ten or twelve yards open into rooms about six yards square. The main drive runs nearly south, and the principal heading strikes east; the rooms take a southerly direction into the centre of the hill. The seam is a splendid one. Its full thickness is seven feet, but a small band of stone runs through it at a height of four feet from the base. The top layer has not been touched; only the four foot seam is worked. It dips almost imperceptibly into the hill. The ventilation of the mine, is effected by an air shaft sunk into it from the side of the hill. The coal is got out in small waggons, which run on a tramway of hardwood and at every corner is a turn-table, also of hardwood.[23] The coal is brought into town by drays, and shipped for the Brisbane market, where it meets with an immediate sale. There are now four miners at work, and the quantity of coal taken out is a little more than 50 tons per week. Two two-ton drays are constantly employed conveying it into town, where a little is disposed of, but most of it is shipped to the metropolis.[24] At a depth of 80 feet, under the present worked seam, is another which crops out at the foot of the hill, and appears to be an extensive one. This has not yet been touched. The Bandambah coal is hard and good, well adapted for steaming or gas purposes, and for transport by ship. The great drawback at this mine is the expense of carting the coal into town. The distance from the river is about two miles, and until a tramway is laid down in this direction, and proper communication established, operations must be on a small scale. An endeavour was made to get up a company to carry on the work extensively, but it failed, owing to the sparcity of money. There can be no doubt that the investment would be a safe and profitable one. The land was taken up under the government regulations.[25]

The coal from the Aberdare Seam is now known to be some of the best in the West Moreton field, particularly for steam raising. This was soon discovered by the Australasian Steam Navigation Company who carried out tests in their ships shortly after operations commenced and declared it to be the equal of Newcastle coal.[26]

By June 1867 Thomas and Thompson had fulfilled the terms of the coal mining lease and Thompson completed the purchase of the Portion 270 Goodna.[27] But his interest in mining flagged and by September 1867 Thomas was advertising that he was carrying on alone at Bundamba although orders could still be sent through the Ipswich office of Thompson,[28] who presumably maintained a financial stake in the business. Competition from other mines was particularly keen in 1868 and Thomas had such a great struggle to keep going[29] that he used to say it was a period of his life when a paid man in his employ was better off than himself.[30] Nevertheless there were encouraging reports of continuing progress[31] and at the end of the year Thomas was able to wrest from William Davies on Sandy Creek the contract for coal for the Ipswich locomotive shops.[32]

Thompson's increasing involvement in politics left him little time for the coal mining venture which relied more and more on the single-handed efforts of Lewis Thomas for its success. In early 1870 the partners jointly purchased Portion 300 Goodna,[33] which adjoined Portion 270 Goodna, but by this time the partnership was under such heavy strain that Thomas decided to close down Bundamba Mine and begin on his own elsewhere.[34] The final closure probably occurred towards the end of 1870. There are unsubstantiated reports that a quarrel between the partners precipitated this action but it can hardly have been very serious because Thompson offered Thomas a half share in the valuable Portion 270 Goodna shortly after Bundamba Mine had closed and the offer was accepted.[35]

Aberdare Mine

During the later stages of his management of Bundamba Mine, Thomas began looking for an opportunity to strike out on his own. By this time he had had considerable experience of mining south of the Bremer and Brisbane rivers, and in 1870 he took out coal mining leases on portions 267, 270, 271 and 272 Goodna[36] which lay immediately to the west of the land on which Redhill Mine had stood and of which he had no doubt heard during his employment with Campbell and Towns (see figure 7 on page 132).[37] He moved into a house on the Brisbane road about two kilometres from the leased land[38] and in September 1870 opened his own mine—the Aberdare[39]—on the lease.[40] The operation was described as follows by a reporter of the *Queensland Times:*

The mine is situated about five miles from town, being about a mile distant to the right of the Brisbane Road . . . Mr. Thomas commenced work by sinking shafts, in the first of which, at a depth of about fifteen feet, he came upon coal. Another shaft was sunk to a depth of about forty feet, at a distance of about two chains farther up the hill, and in this excellent coal was obtained within a few feet of the surface. A mouth was then opened from the surface at the first sinking, and a tunnel was driven for about a hundred yards to the dip of the coal, which runs due south. The drive is carried down about six yards wide, with a horse-tramway on one side . . . The seam is four feet thick, and the coal has the appearance of being of the anthracite description, being fine, hard, brittle, and sparkling, and entirely free from any earthy matter. The bed of the coal is intersected with a hard brown flinty stone stratum, of about an inch thick, and occurring at intervals of some eighteen inches throughout. This necessitates considerable labour to keep the coal free from stone, but as it comes off in fair-sized patches it can be readily picked from the coal, which latter is then trucked away in its purity, the flinty stone being left behind in the mine. The wooden tramway to which reference has already been made, runs the whole length of the main tunnel, and the trucks containing the quarried coal are placed on a carriage-way two at a time, and driven away by a horse to small shoots erected at the opening of the pit for the purpose of receiving the coal, and from which drays are laden to carry it away to its destination. About seventy yards down the main drive a stage is erected to receive the trucks as they are filled by the miners at work in the several chambers. This stage was erected for the purpose of saving the construction of turn-tables, and the necessity of taking a horse into all the chambers, as was at first intended, and it has been found to answer the purpose exceedingly well, as there is no time lost either by the miners waiting for the empty trucks, or by the drayman, who takes the full ones away, leaving the empty ones in their place.

Parallel with the main tunnel, a second drive has been opened, about ten yards wide and four feet high, which is worked from the forty feet shaft, two chains higher up than the original opening. This tunnel has been worked the entire length of the main drive, and serves to ventilate the mine. The air is really excellent, being pure and cold, and as yet the miners have not had much trouble as regards dampness or water in the mine.

The lateral working of the mine has been commenced by a cross-heading about 70 yards down the main tunnel, where the miners are at present engaged. It has been carried back about 50 yards at right angles, and will ultimately extend as far as the boundary of the property. From this heading three chambers, each nine yards wide, are already opened, the purpose being to work these in galleries towards the rise of the coal, leaving pillars two yards square as a support for the roof, which as well as the floor is of pure coal. In every direction in which the mine has as yet been worked, coal has been found, and generally, running through the seams, are the strata of flint stone already referred to. The propping of the tunnel, and of the chambers at present in working, has been effected by logging at distances of about six feet apart. . .

The labour was as a matter of necessity limited at the start, but there are now nine miners constantly engaged in the pit, exclusive of the hands who remove the coal to the minor shoots. Larger shoots have been erected within the last two months at the Chestnut Scrub, on the bank of the Brisbane River, a short way below its junction with the Bremer, and about two and a half miles from the mine.[41] They contain about sixty tons of coal, several drays being constantly employed between these shoots and the mine . . .

From twelve to fifteen tons per day are taken from the mine, and demand is as great as the present capacities of the establishment can supply.[42]

Figure 8 shows a reconstruction of the underground plan of Aberdare Mine.

Figure 8. Reconstruction of the plan of the workings of Lewis Thomas's Aberdare Mine at Dinmore. Solid lines enclose parts mapped when the mine was uncovered during recent strip-mining operations. Heavy broken lines identify portions described by a visitor to the mine in 1871 (chap. 10, n. 42). Light broken lines indicate parts reconstructed on the basis of contemporary mining practice. See fig. 10, p. 162.

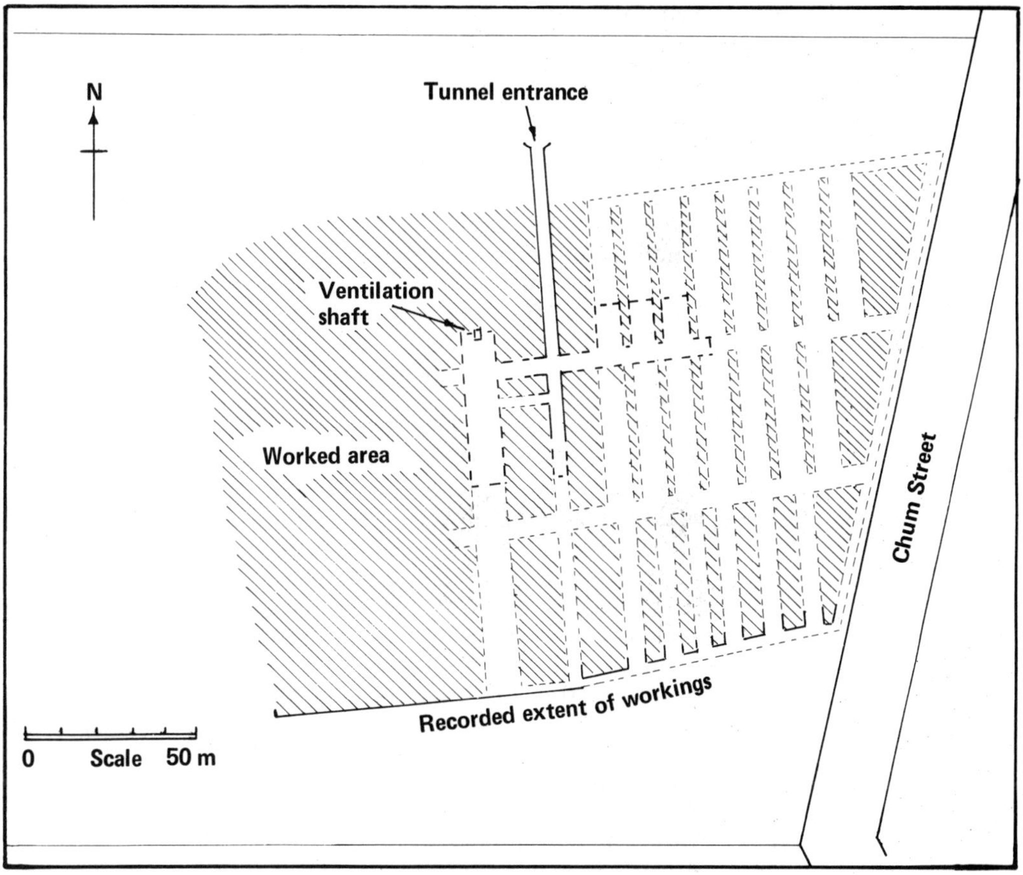

The Aberdare Mine at Dinmore worked the Bluff Seam[43] whose quality, although acceptable for some purposes, was inferior to the Bundamba coal in most respects, ash assays of 24.2 per cent[44] and 14.0 per cent[45] being quoted by contemporary writers. It was also difficult to mine because the seam dipped steeply.[46] Nevertheless Thomas completed the purchase of the portions in 1872[47] and by 1873 had advanced the main drive to 125 yards (112 metres),[48] but in 1874 he failed in an attempt to capture the contract for fuel for the locomotives of the Southern and Western Railway which was awarded to the Old Tivoli Pit.[49] Production at Dinmore failed to increase and when Thomas finally succeeded in gaining full control of portions 270 and 300 Goodna from Thompson in 1874,[50] the way was clear for him to return to Bundamba and resume operations there. In 1876 he opened a mine close to the old Bundamba Mine and in 1877 closed down Aberdare Mine at Dinmore.[51] Thomas's days of apprenticeship and uncertainty were at an end. The coal in the new mine was of excellent quality and occurred in large quantities so that mining could be expanded without difficulty. During the next thirty years Thomas led the West Moreton coalfield to an unprecedented expansion which brought prosperity to the region and made him the first Queensland coal magnate.

Of the early operations south of the Bremer River very little now remains. Campbell and Towns's pioneer effort at Redhill has been completely swallowed up in later open-cut operations, and the same is true of Lewis Thomas's Aberdare Mine.

Although the valley in which Thompson and Thomas's Bundamba Mine was situated has not been open-cut the area has been extensively reworked at various times. The 1899 geological survey map[52] marks the position of some seven tunnel entrances driven into the sides of the Blackstone Valley without indicating which of them (if any) was Bundamba Mine. From the contemporary description of the operation quoted earlier[53] it appears that the opening was located on the southern flank. The same article also mentions that the principal heading was driven to the east so that one might reasonably infer that the main tunnel was situated towards the western end of the valley. In view of the frequently-reported difficulties of hauling coal to Ipswich it would clearly have been advantageous to minimize this distance by placing the first mine at the western end of the valley where the road from Ipswich entered it. Support for this view can be found in the article mentioned earlier which stated that the entrance was "... at a bight where two hills meet". The Bundamba Mine has therefore been indicated in plate 37 as the most south westerly working in the Blackstone Valley of those shown on the 1899 geological survey map. At this location a ground inspection showed very clear signs of the collapse and slippage of the sandstone strata which are known to have overlain the Aberdare Seam.[54] No sign has been found of the coal loading wharf which Thomas constructed downstream of the Moggill ferry at Chestnut Scrub to ship the coal from his Aberdare Mine.

Plate 37. Blackstone, showing the possible site of Bundamba Mine. (Photographed in May 1979.)

Notes and References for Chapter 10

1. *Queensland Times*, 28 October 1871.
2. Queensland Government Department of Mapping and Surveying, Brisbane, Map S31.85.
3. Queensland Government Titles Office, Vol.116, Folios 209 and 210, 12 July 1867; *Queensland Times*, 20 September 1866; and 8 September 1866.
4. *Towns* v. *J. and J.E. Campbell*, 1867. Queensland State Archives. Accession 25/92. SCT.38 (John Corbett's Testimony); and (James Robertson's Testimony).
5. *Votes and Proceedings of the Legislative Assembly of Queensland*, 1876, Vol.3, pp.387-95.
6. *Australian Dictionary of Biography*, 1851-1890, Vol.6 (Carlton: Melbourne University Press, 1976), pp.260-61.

7. M. Fox, *The History of Queensland*, Vol.1 (Brisbane: State Publishing Co.), p.447.
8. *Australian Dictionary of Biography*, 1851-1890, Vol.6, pp.260-61.
9. *Votes and Proceedings of the Legislative Assembly of Queensland*, 1867, Vol.2, p.200.
10. *Queensland Times*, 17 August 1865.
11. Ibid., 17 April 1866.
12. Ibid., 17 February 1913.
13. *Australian Dictionary of Biography*, 1851-1890, Vol.6, pp.265-66.
14. D.B. Waterson, *A Biographical Register of the Queensland Parliament, 1860-1929* (Canberra: Australian National University Press, 1972), p.178.
15. *North Australian*, 29 December 1857.
16. *Moreton Bay Courier*, 14 June 1861.
17. A variety of spellings has been used in the past for "Bundamba". The currently-accepted spelling is used uniformly here in order to avoid confusion. Exceptions are made only in cases of direct quotations from contemporary documents when the original spelling is retained.
18. Queensland Government Titles Office, Vol.116, Folio 127, 29 July 1867.
19. R. Spencer Brown, *A Journalist's Memories* (Private Publication, 1927), p.82.
20. *Australian Dictionary of Biography*, 1851-1890, Vol.6, pp.265-66.
21. *Votes and Proceedings of the Legislative Assembly of Queensland*, 1865, pp.1207-8.
22. *Queensland Times*, 17 July 1866.
23. The gauge of the tramway was 2 feet (0.6 metres). *Queensland Times*, 13 September 1866.
24. Coal shoots were erected at the end of Thorn Street in Ipswich to load the coal into lighters for transport down river.
25. *Queensland Times*, 20 December 1866.
26. Ibid., 13 September 1866.
27. Queensland Government Titles Office, Vol.116, Folio 127, 29 July 1867.
28. *Queensland Times*, 19 January 1867.
29. Ibid., 17 February 1913.
30. *Pugh's Almanac* (Brisbane: Thorne & Greenwell, 1894), p.149A.
31. *Queensland Times*, 16 May 1868.
32. *Queensland Government Gazette*, 1869, Vol.10, p.47.
33. Queensland Government Titles Office, Vol.162, Folio 92, 2 June 1870.
34. *Votes and Proceedings of the Legislative Assembly of Queensland*, 1871, pp.675-837.
35. Queensland Government Titles Office, Vol.116, Folio 127, 14 August 1871, Transfer 25643.
36. Queensland Government Titles Office, Vol.188, Folio 75, 17 April 1872.
37. Waterson, *Register of the Queensland Parliament*, pp.177-78.
38. *Queensland Times*, 28 October 1871.
39. It is not unusual in the West Moreton coalfield for a colliery and the main or original seam which it worked to share the same name; Tivoli and Aberdare are examples. Additional confusion arises in the case of Aberdare by two quite different collieries bearing this name. When Thomas and Thompson commenced operations on Portion 270 Goodna adjacent to Bundamba Creek the mine was named in contemporary writings as Bundamba Mine but the seam was unnamed. On moving to Dinmore to begin his own operation on portions 267, 270, 271 and 272 Goodna (the duplication of Portion 270 Goodna should be noted) Thomas decided to call it Aberdare Mine (figure 7). However, when he later returned to the original

Portion 270 Goodna at Bundamba he again called his mine the Aberdare and the main seam came to be known by the same name. The mine and the coal seam which he had worked at Dinmore were then termed the Old Aberdare. See W.E. Cameron, "Geology of the West Moreton or Ipswich Coalfield", *Queensland Geological Survey*, Publication 147 (Brisbane: Government Printer, 1899).

40. *Australian Dictionary of Biography*, 1851-1890, Vol.6, pp.260-61; and pp.265-66.
41. Thomas had bought from John Parton subdivisions 29 and 36 of Portion 131 Goodna which were situated immediately downstream of the present Moggill ferry. The wharf is shown on a contemporary map (Queensland Government Department of Mapping and Surveying, Maps M31.36 and M33.137) as being near the north boundary of subdivision 29 (figure 7).
42. *Queensland Times*, 28 October 1871.
43. A.K. Denmead, "The West Moreton (Ipswich) Coalfield", *Queensland Geological Survey*, Publication 279 (Brisbane: Government Printer, 1955).
44. R. Daintree, *Queensland, Australia* (London: G. Street, 1873), p.78.
45. *Votes and Proceedings of the Legislative Assembly of Queensland*, 1876, Vol.3, pp.387-95.
46. Ibid.
47. Queensland Government Titles Office, Vol.188, Folio 75, 17 April 1872.
48. Daintree, *Queensland, Australia*, p.78.
49. *Votes and Proceedings of the Legislative Assembly of Queensland*, 1876, Vol.3, pp.681-706.
50. Queensland Government Titles Office, Vol.178, Folio 151, 26 January 1874; and Vol.162, Folio 92, 21 November 1873, Transfer 32102.
51. *Votes and Proceedings of the Legislative Assembly of Queensland*, 1878, Vol.2, pp.319-402.
52. Cameron, "Geology of the West Moreton or Ipswich Coalfield".
53. *Queensland Times*, 28 October 1871.
54. Denmead, "The West Moreton (Ipswich) Coalfield".

Coal Mining outside Ipswich

11

The Mary River Mines

The discovery of coal in the Maryborough region probably dates from 1863, in October of which year a Mr W.C.G. Giles called at the offices of the *Maryborough Chronicle* and left ". . . a very fine sample of coal obtained in the neighbourhood of his farm on the banks of the Mary".[1] The sample seems to have come from the Upper Mary, some twenty-four miles from the Custom House and near the present site of the township of Tiaro. A year later a forty-six year old Welsh immigrant, Abel Staley, opened a small mine in the area but he soon disposed of it to the Mary River Coal Company, and moved to the Burrum River where very promising finds had recently been reported. The pit on the Mary River struggled on for a further year under new management despite the intractable problems of transporting coal down a shallow, twisting river and working a poor seam less than 3 feet (0.9 metres) thick.[2] James Robertson, a Scottish miner recently arrived in Queensland,[3] who was later to give evidence in the law suit between Towns and the Campbells over the Redbank Mine, and then to establish important mining interests on the Burrum field, was highly critical of the area, commenting that ". . . the rocks are much troubled and the coal is foul, or nearly anthracite . . . ".[4] Mining finally ceased in mid 1866 when the workings became flooded with water and silt,[5] but their exact location still remains unknown.

The hazards of shipping coal down the dangerous, winding Mary River led local inhabitants to seek deposits nearer the mouth and in 1868 a former Maryborough town surveyor, Mr Richard J. Ramsay, reported the discovery of a suitable site some ten kilometres downstream from the town, conveniently near to the left bank and close to Saltwater Creek. Ramsay proceeded to sink a shaft which cut through a narrow band of coal after 66 feet (20 metres) but had not reached anything worthwhile after being excavated for a further 54 feet (16 metres). At this depth his finances failed and as he was unable to timber the shaft the men refused to enter it because of its insecurity. Ramsay therefore filled in the bottom 50 feet (15 metres), worked the one narrow seam as best he could and prevailed upon Governor Blackall to christen his endeavour the Durisdeer Coal Mine during his visit to the area in September 1869. Shortly afterwards Ramsay solved his financial problems, temporarily at least, by floating the operation as a limited company—Durisdeer Coal Mining Company Limited—with a capital of £5,000 ($10,000) and himself as a director and engineer-in-charge. In the company prospectus the point was made that the proprietors had prospected for coal on the site by boring and had found a seam 4 feet (1.2 metres) thick at a depth of 176 feet (53 metres) situated about a kilometre from the river. Shaft-sinking commenced at that spot early in 1870 and by July it was down to 100 feet (30 metres), although the inflow of water was becoming difficult to control. A 7 inch (175 millimetre) diameter pump, driven by a 10 horsepower (7.5 kilowatt) engine, was ordered from the newly established company of John Walker and Company of Maryborough and by September shaft sinking was proceeding again. After advancing a further 38 feet (11.4 metres) a borehole was driven ahead of the shaft to confirm the depth and thickness of the seam, but the result was disturbing in that coal was only reached at a depth of 263 feet (79 metres) from the surface, instead of the 176 feet (53 metres) as previously predicted. The directors sought a second opinion on the quality of the seam and invited James Robertson, who had recently joined the board of the company, to inspect the cuttings as the drill bored through the seam.[6]

Robertson was born in Scotland and came from a family whose mining traditions reached back through his father to his grandfather. In 1853 when a young man he joined his brother in Glasgow to practise as a mining engineer, surveying and reporting on the mines of the Clyde Iron Company which were situated in the Kilmarnockshire and Ayrshire coalfields. After some four years he leased and ran his own mine in Scotland before giving it up to emigrate to Queensland in 1865.[7] In 1868 when he was appearing for John Campbell in the Redbank Mine dispute he referred to himself as a "duly qualified Mining Engineer",[8] but his qualifications do not appear to have included any formal certification.

Robertson pronounced the quality of the coal to be good although there was an ominous addendum to the effect that it included "... slight ribs of earth matter".[9] Unfortunately by this time the company's finances were exhausted but by issuing more shares, and inducing John Walker and Company to accept some of them in payment for the pump and engine, the directors were able to muster sufficient funds to call tenders for the sinking of the final 130 feet (39 metres) of the shaft. In April 1871 the contract was awarded to a Mr H. Mackenzie and the directors imagined that the worst was behind them.[10] But then a series of minor disasters overtook the company.

Hardly had the contractor started sinking the shaft than he was set upon by Aborigines on his way to work and seriously injured. Desperately anxious to reach some coal, James Robertson, who had by this time taken over from Richard Ramsay as engineer, persuaded his fellow directors to dewater, clean out and complete the old shaft the short distance to the coal seam. A trial boring proved that the supposed "few feet" requiring penetration were in fact 40 feet (12 metres). Moreover, the contractor who carried out the work, successfully sued the company for a considerably larger sum than the directors considered his entitlement. Once again the directors faced a financial crisis but this time there was no alternative to liquidation and the company was wound up voluntarily in December 1871. Subsequent attempts by other parties to reach the coal supposedly lying below both of the shafts were unsuccessful and the site was ultimately abandoned.[11]

A later judgment was that the operation failed because the shafts were sunk on a fault and met with intractable flooding problems.[12] Whether the coal would have been of saleable quality if it had ever been reached is unknown.

The Burrum Coalfield

During the first half of the 1860s Maryborough developed quickly and by 1865 sugar and timber were being shipped regularly to Brisbane through its port[13] while two years later John Walker and Company began supplying local engineering needs.[14] The coastal steamers of the Australasian Steam Navigation Company and the Queensland Steam Navigation Company called fairly frequently and the time was ripe for the establishment of a secure supply of local coal to support the various enterprises.

The country lying to the north west of Maryborough is remarkable for its flatness and lack of surface features, even the watersheds being barely recognizable under a thick cover of stunted wattle, grass trees, ti-trees, honeysuckle and bloodwood.[15] The channel of the Burrum River cuts a steep-sided twisting course through the scrub-covered plain, exposing outcrops of coal in its sandstone banks for a distance of some three miles downstream from the crossing of the Maryborough to Bundaberg road. It was in this area in May 1863 that Robert Miller in company with his brother John found coal.[16] Claims were made at a later date that the Millers only rediscovered what had already been found and that the presence of coal on the Burrum River was recognized before Separation by a local licensed surveyor, Mr Buchanan, but there is no real evidence of this. In any case there appeared to be no obvious market for it in 1863 and the Miller brothers put their energies into establishing a timber business.[17]

In January 1865 the owner of a nearby sawmill, Mr Greathead, approached Robert Miller and suggested that the steam-raising capabilities of the local coal be tested under the boilers in his mill.[18] A small supply was procured from the banks of the Burrum River and the results were so encouraging that Miller cut and shipped further samples to Maryborough by water[19] for testing by other consumers.[20] These included the Australasian Steam Navigation Company who experimented with it in their boats,[21] and the relatively newly-formed Queensland Steam Navigation Company who tried it in their ship the *Queensland*.[22] The reactions were unanimously favourable and Miller immediately decided to expand his business.[23]

The Burrum coalfield forms an elongated basin twenty kilometres in length and ten kilometres in width with its major axis running from north west to south east.[24] It contains six workable seams extending over a total depth of about three hundred metres which outcrop only in the area where the channel of the Burrum River cuts through the rim of the basin. At this point the seams dip steeply under sandstone cover, the most conspicuous being the Burrum (1 metre thick at the outcrop), Watson's (1.2 metres thick) and Lapham's (0.75 metres)[25] (figure 9). An ash assay on a coal sample taken from the outcropping Burrum seam in 1878 yielded the comparatively low ash assay of 8 per cent[26] and it was this seam which was the first to be worked.[27]

In 1865 the Burrum seam was visible in both banks of the river, the face of the seam being exposed on the left-hand bank for a length of about 30 metres and a thickness of 1.1 metres just above the level of the river. On the right-hand bank the seam was so near to the water that it became inaccessible at high tide.[28] Miller commenced his operations on government land on the right-hand bank but then switched to the left-hand,[29] and fossicked on both of them according to the state of the river,[30] aiming at extracting some twenty tonnes a month in total.[31]

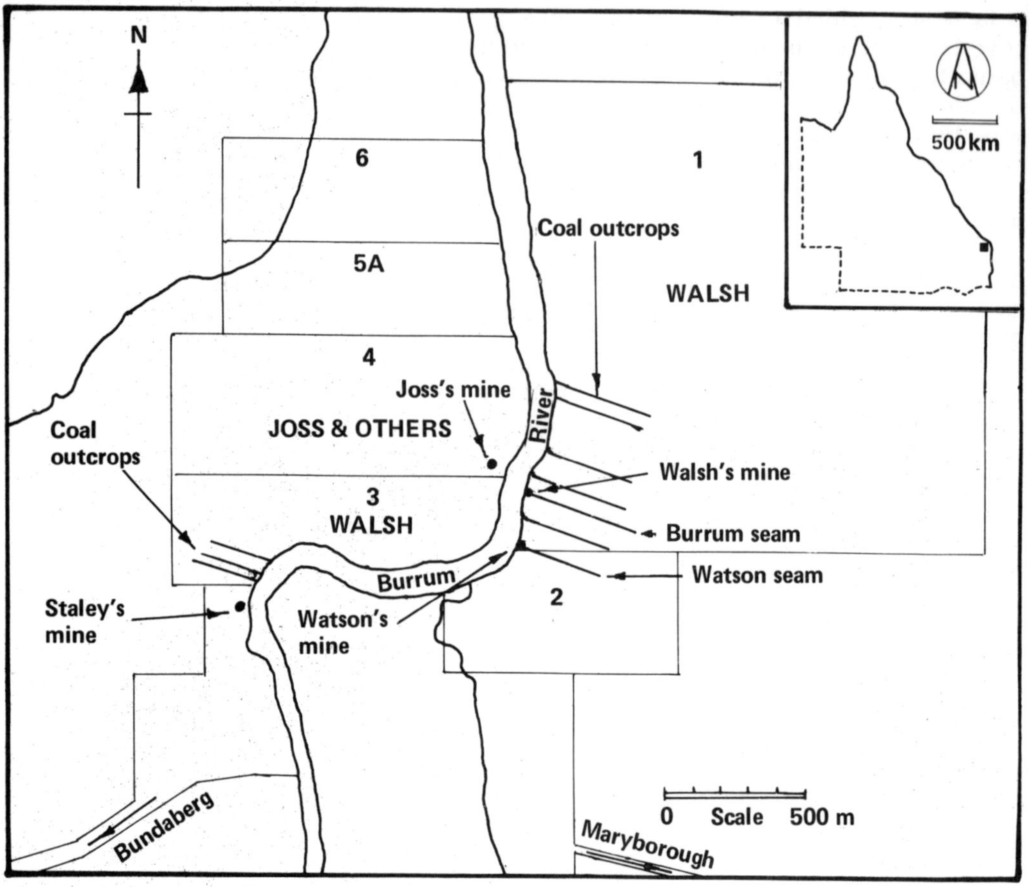

Figure 9. The Burrum area showing original subdivisions and early coal mines. (After "The Burrum Coal Field", *Queensland Government Survey*, Publication 170, 1902.)

Meanwhile an impecunious Newcastle miner, William Morris, had heard of the discovery of coal on the Burrum River and he journeyed north to inspect it.[32] Miller talked freely to Morris about his finds but indicated that capital was required to develop them.[33] The visitor kept his own counsel and returned to Brisbane where, with the support of two equally impecunious friends, James Patterson and John Cameron, he set to work to acquire the Burrum coal-bearing land from the government.

The first move by the partners was to present themselves to the surveyor-general and make application for a lease over the area,[34] but this was refused because at that time there were no regulations for leasing lands for coal mining. Their next approach was to a "gauche, nasty, devious, highly egocentric" politician[35] William H. Walsh, the newly elected MLA for Maryborough, who had already received substantial land grants in the area[36] and who had influential contacts in the government (plate 38). Walsh persuaded the premier that regulations for leasing or granting coal-bearing lands were urgently required in

Plate 38. William H. Walsh. (Courtesy John Oxley Library.)

Queensland if production were to be expanded. The premier agreed to introduce appropriate legislation and Walsh advised the conspirators to look out for it. They, however, had neither the experience nor the funds to make an application and Walsh therefore agreed to apply on their behalf [37] on the understanding that a partnership be formed of which he would hold one half. The trio willingly accepted this proposition.

Meanwhile Miller recognized the dangerous position in which his indiscreet disclosures had placed the security of his coal foraging operation and on 14 August he wrote to the government, applying for the purchase or lease of the land where he was mining.[38] Unfortunately he omitted to specify any particular area and no action was taken in Brisbane. Three days later Walsh submitted a claim to a square mile (2.5 square kilometres) of land (the largest area allowable) on the right bank of the Burrum River where Morris had indicated that the coal was located.[39] He paid the required deposit of £160 ($320) and set off to Maryborough to inspect his claim. But the information which Morris had given Walsh was incorrect and the claimed area contained only a small part of the outcrop on the river. The local surveyor was persuaded to redraw the boundaries and Miller's best coal-gathering area was swept into it.[40] The new legislation was promulgated on 31 August 1865[41] and Walsh had notices put in the local paper warning trespassers from removing coal from the land of what was termed the Burrum Coal Company.[42] Reacting angrily to this intrusion, Miller immediately set out for Brisbane, accompanied by another local coal prospector, John Falconer,[43] and on 16 September he lodged an application for 80 acres (32 hectares) (the smallest allowable lease) which covered his little mine.[44] His friend made an application for 130 acres (52 hectares) on the opposite bank in the name of a combine of local men headed by James Joss[45] which was accepted, but Miller's application was refused because the land had already been assigned to Walsh and his partners.

Miller returned north and resolutely continued working his mine while loudly appealing through the press for justice[46] and petitioning the parliament in Brisbane.[47] But his cry went unheard. At the end of the month Morris and Patterson arrived on the river and after trying to evict him with "violence and threatening language"[48] they finally filled in the tunnel entrance and forced him to leave. As soon as he had gone the partners set to work under the direction of Morris[49] to reopen and improve the operation.[50] The entrance situated about 6 feet (1.8 metres) above the high-water mark led to a main drive 6 feet high (1.8 metres) and 7½ feet (2.25 metres) wide which extended for 50 yards (45 metres) and was serviced by an iron-railed tramway. A ventilation shaft 40 feet (12 metres) in depth was sunk and a heading 60 yards (54 metres) long driven from the main tunnel. A few rooms were opened, the number of employees raised to nine, and by February 1866 there were optimistic reports that the proprietors were looking forward to delivering a hundred tonnes of coal daily.[51]

Walsh, however, began to regret the partnership into which he had so eagerly bought.[52] The cost of mining the coal was much greater than the selling price, the competence and industry of his partners was, to say the least, untried and he had listened to subsequent advice which indicated that it would have been safer to have started a shaft operation than to have continued tunnel mining. When his partners refused to accept this advice Walsh played his trump card. Although the original agreement had been that Walsh would acquire land on behalf of the partnership he had actually made the application in his own name.[53] At the end of February, therefore, he obtained an injunction restraining Morris, Patterson and Cameron from working on the land and he closed the mine although Morris defiantly continued removing and selling the stock coal which had already been cut.[54] Whether in the first year an output of 640 tonnes was reached in order to satisfy the terms of the government lease is unknown but it is likely that it was, because Walsh's ownership of the land was subsequently confirmed. A visitor to the area in October 1866 reported that the mine was at a standstill[55] and it is unlikely that it was ever reopened.

Meanwhile on the left bank of the river and slightly downstream from Walsh's mine the small syndicate led by James Joss set to work sinking a shaft close to the water[56] on the land which they had leased. By October 1866 a steam engine had been installed for hoisting and pumping duties and a thousand tonnes of coal had been extracted.[57] But the partnership was short-lived and production began to tail off. In early 1868 the mine was put up for sale[58] and all operations appear to have ceased.

Abel Staley, the pioneer of coal mining on the Upper Mary River, had also been attracted to the Burrum River and in mid 1866 he opened a mine on the left-hand bank a short distance above the ill-fated Walsh tunnel mine. This provided coal for the Australasian Steam Navigation Company's steamers at Maryborough during 1868 and survived floods on the river in 1869 before succumbing to even heavier floods in 1870. Staley abandoned coal mining and went farming, although this did not deter him from dabbling intermittently in developments on the Burrum River over the following decade, until his death in 1880 finally robbed the area of one of its most resolute characters.[59]

In 1870 a fourth attempt was made to establish a viable operation on the river, this time by an immigrant miner from England, Thomas Watson. But he was even less successful than his predecessors and by the end of the year all mining had ceased. Other attempts were similarly unrewarded.[60] The highest annual production figures for the Burrum River coalfield during this period were reached in its first year in 1866 when the joint output of the three mines of Walsh, Joss and Staley approached two thousand tonnes, although Staley on his own managed to exceed one thousand tonnes in 1869 before being overwhelmed by floods.[61] Otherwise the output was extremely small and intermittent.

There were two obstacles to maintaining a sustained coal production. The first was that the seams only outcropped on the banks of the river where they were barely above the normal water level. This made any tunnel works extremely vulnerable to inundation in wet weather and at high tides. Moreover the water forced its way between the bedding planes in the seam, leading to weathering and oxidation of the coal and a deterioration in quality,[62] although this did not deter groups and individuals (which included James Robertson) from taking up conditional grants of land for mining coal elsewhere along the Burrum River and in its vicinity.[63] In practically all cases no outcrop coal was visible to guide the lessees, who had no alternative to sinking shafts or boring holes to find the hidden seams. This was an expensive and risky business which proceeded slowly and fitfully.

The second obstacle was the expense and difficulty of carting the coal to Maryborough. The road was unable to carry continual heavy loads of coal and the alternative route by water necessitated a haul of sixteen miles (twenty-five kilometres) down the Burrum River followed by forty miles (sixty-four kilometres) across Wide Bay and a further sixteen miles (twenty-five kilometres) up the Mary River to Maryborough. The journey often took a week owing to the adverse influence of the tides and the presence of numerous sand banks in the Burrum River.[64] A sustained output had to await the completion of the railway line from Maryborough to Burrum in 1883[65] and the sinking of shafts well back from the banks of the river.[66]

The only one of these small operations which has been identified is Walsh's mine on the bank of the Burrum River. Its position was located by Mr David Bailey in 1971 who partially unearthed the entrance and photographed it (plate 39).

Plate 39. Entrance to Walsh's mine in 1971, showing the timbers supporting the portal. (Courtesy David Bailey.)

Darling Downs

The presence of coal in large quantities on the Darling Downs was reported in 1853 by the Rev. W.B. Clarke who observed that they were of a cannel rather than a bituminous type.[67] The opening of the railway line from Toowoomba to Allora in March 1869 and its extension to Warwick nearly two years later[68] sparked off an interest in assessing these deposits with the object of exploiting them, particularly where they were situated close to the line.

Following the sinking of an experimental shaft some 40 feet (12 metres) deep in 1869,[69] a couple of small mines were opened at Allora about three kilometres east of Hendon railway station.[70] Support for the ventures came from the Southern and Western Railway who in 1871 awarded a contract to Messrs Shortall, Monroe and Company for delivering coal to the railway at Allora[71] following the conversion of locomotive firing from wood to coal.[72] The ash content of the coal was claimed to be only 8.9 per cent[73] but the seam was less than 2 feet (0.6 metres) thick[74] and tests on a few hundred tonnes of it[75] suggested that the coal might be more suitable for gas making than steam raising.[76] Unfortunately, coal from the Darling Downs has always suffered from a common characteristic of cannel coals. This is that they seldom form a satisfactory coke even though they are prolific gas producers, and their inadequate coking performance was the reason given by the gas works in Brisbane for not using them.[77] The works continued to draw most of its coal from Newcastle, leaving the proprietors of the Allora mines to struggle on as best they could until 1875, when they gave up the fight and closed down their operations.

Notes and References for Chapter 11.

1. *Maryborough Chronicle*, 15 October 1863.
2. Ibid., 19 July 1865; 6 May 1880; 23 October 1879; 28 January 1865; 11 March 1865; 22 March 1865; 11 January 1865; 22 March 1865; 12 November 1864; and 24 June 1865.
3. *Votes and Proceedings of the Legislative Assembly of Queensland*, 1900, Vol.2, pp.441-42.
4. *Maryborough Chronicle*, 9 September 1865.
5. Ibid., 7 July 1866.
6. Ibid., 19 May 1868; 9 February 1869; 9 September 1869; 1 January 1870; 30 September 1869; 6 August 1870; 30 September 1869; 23 April 1870; 6 August 1870; 1 October 1870; 17 January 1871; 26 November 1870; 20 December 1870; 6 August 1870; and 20 December 1870.
7. *Votes and Proceedings of the Legislative Assembly of Queensland*, 1900, Vol.2, pp.441-42.
8. *Towns* v. *J. and J.E. Campbell*, 1867. Queensland State Archives. Accession 25/92. SCT.38 (James Robertson's Testimony).
9. *Maryborough Chronicle*, 17 January 1871.
10. Ibid., 11 February 1871; 18 February 1871; and 1 April 1871.
11. Ibid., 4 May 1871; 9 September 1871; 31 October 1871; 9 February

1869; 23 November 1871; 7 December 1871; and 19 September 1878.

12. Ibid., 19 September 1878.
13. N.L. McKellar, *From Derby Round to Burketown* (Brisbane: Queensland University Press, 1977), p.35.
14. *Maryborough Chronicle*, 6 August 1870; and J. Armstrong, "Walkers Ltd.", *Australian Historical Railway Society Bulletin*, Vol.19, 1968 (New Series), pp.229-32.
15. *Votes and Proceedings of the Legislative Assembly of Queensland*, First Session, 1879, pp.511-12.
16. *Queensland Daily Gazette*, 7 October 1865.
17. *Maryborough Chronicle*, 11 November 1865.
18. *Queensland Daily Gazette*, 7 October 1865.
19. *Maryborough Chronicle*, 28 January 1865; and 21 June 1865.
20. *Votes and Proceedings of the Legislative Assembly of Queensland*, 1866, p.1473.
21. *Maryborough Chronicle*, 21 June 1865; and *Brisbane Courier*, 28 July 1865.
22. *Brisbane Courier*, 29 August 1865.
23. *Maryborough Chronicle*, 21 June 1865.
24. D.C. Mengel, "Ipswich Coal-Field, Queensland", *Economic Geology of Australia and Papua New Guinea*, Vol.2, ed. D.M. Traves and D. King (Melbourne: Australasian Institute of Mining and Metallurgy, 1975), p.290.
25. *Votes and Proceedings of the Legislative Assembly of Queensland*, First Session, 1879, pp.511-12.
26. Ibid.
27. Queensland Government Department of Mapping and Surveying, Brisbane, Map W37.12; and Map W37.22.
28. *Maryborough Chronicle*, 21 June 1865.
29. *Queensland Daily Gazette*, 7 October 1865.
30. Queensland Parliamentary Debates, 1865, Vol.3, pp.464-74.
31. *Maryborough Chronicle*, 4 November 1865.
32. *Queensland Daily Gazette*, 7 October 1865.
33. *Queensland Times*, 30 September 1884.
34. Queensland Parliamentary Debates, 1865, Vol.3, pp.464-74.
35. *Australian Dictionary of Biography, 1851-1890*, Vol.6 (Carlton: Melbourne University Press, 1976), p.348.
36. Queensland Parliamentary Debates, 1865, Vol.3, pp.464-74.
37. Ibid.
38. *Votes and Proceedings of the Legislative Assembly of Queensland*, 1866, pp.1477-78.
39. *Queensland Daily Gazette*, 7 October 1865.
40. Ibid.
41. *Queensland Government Gazette*, 1865, Vol.6, p.746.
42. *Maryborough Chronicle*, 13 September 1865.
43. *Queensland Daily Gazette*, 10 October 1865.
44. *Votes and Proceedings of the Legislative Assembly of Queensland*, 1866, p.1473.
45. Queensland Government Department of Mapping and Surveying, Map W37.21.
46. *Queensland Daily Gazette*, 14 November 1865; and 7 October 1865.
47. *Votes and Proceedings of the Legislative Assembly of Queensland*, 1866, p.147.
48. *Queensland Daily Gazette*, 14 November 1865.
49. *Towns* v. *J. and J.E. Campbell*, 1867. Accession 25/92. SCT.38 (Robertson).
50. *Queensland Daily Gazette*, 3 February 1866.
51. *Maryborough Chronicle*, 9 December 1865; *Queensland Daily Gazette*, 5 June 1866; and 3 February 1866.
52. Queensland Parliamentary Debates, 1865, Vol.3, pp.464-74.
53. *Votes and Proceedings of the Legislative Assembly of Queensland*, 1866, p.1477.

54. *Queensland Daily Gazette*, 1 March 1866; and 19 June 1866.
55. *Maryborough Chronicle*, 31 October 1866.
56. Queensland Government Department of Mapping and Surveying, Map W37.21.
57. *Maryborough Chronicle*, 31 October 1866.
58. Ibid., 12 February 1868.
59. Ibid., 11 January 1865; 6 June 1868; 23 February 1869; 5 February 1870; 19 February 1870; and 11 April 1871; and *Queenslander*, 15 May 1880.
60. *Maryborough Chronicle*, 11 April 1871; 4 January 1872; 15 May 1873; and 17 June 1873.
61. *Statistics of Queensland*, 1869.
62. *Votes and Proceedings of the Legislative Assembly of Queensland*, First Session, 1879, pp.511-12.
63. *Maryborough Chronicle*, 26 November 1874.
64. *Votes and Proceedings of the Legislative Assembly of Queensland*, First Session, 1879, pp.511-12; and 1880, Vol.2, pp.978-81.
65. *Queensland Government Gazette*, 1883, Vol.33, p.395.
66. *Votes and Proceedings of the Legislative Assembly of Queensland*, First Session, 1879, pp.511-12.
67. *Votes and Proceedings of the Legislative Council of New South Wales*, 1853, Vol.2, pp.853-70.
68. *Queensland Government Gazette*, 1869, Vol.10, p.322; and 1870, Vol.11, p.1700.
69. *Queenslander*, 20 November 1869.
70. *Votes and Proceedings of the Legislative Assembly of Queensland*, 1876, Vol.3, pp.387-95; and *Queensland Times*, 18 March 1871.
71. *Queenslander*, 27 May 1871.
72. *Votes and Proceedings of the Legislative Assembly of Queensland*, 1871, pp.661-73.
73. R. Daintree, *Queensland, Australia* (London: G. Street, 1873), pp.77-80.
74. *Votes and Proceedings of the Legislative Assembly of Queensland*, 1876, Vol.3, pp.387-95.
75. *Queenslander*, 17 June 1871.
76. *Votes and Proceedings of the Legislative Assembly of Queensland*, 1871, pp.675-837; and 1878, Vol.2, pp.319-402.
77. Ibid.

Technical Aspects of Early Coal Mining in Queensland

12

There can be no doubt that before the 1880s coal mining in and to the north of the Moreton Bay area was a small, insignificant industry even by contemporary Australian standards. In 1870 after a decade of colonial independence, production in Queensland had barely reached twenty thousand tonnes per annum which was about 3 per cent of the coal being mined in New South Wales at that time. In 1871, the amount of coal produced per head of the population was only 7 per cent of the New South Wales figure,[1] and expansion had been equally slow before Separation despite the efforts of the New South Wales government to foster the expansion of the infant industry by terminating the monopoly coal mining rights which had been unwisely granted to the Australian Agricultural Company. The chief initiative by the government had been to identify certain areas as "coal allotments" and to auction them off in small portions ahead of other land where coal production possibilities were less obvious and presumably more demanding. The potential of the allotments as coal producers was certainly recognized by the local residents who took them up enthusiastically as they became available. But the buyers then preferred to sit on their purchases and do nothing instead of initiating some action. Stutchbury pointed out

> . . . that parties in the neighbourhood have not neglected the opportunity of purchasing land where coal "cropt" out, thus securing to themselves the most available spots where it could be immediately obtained, yet, considering the demand and the frequent delay, even to their own steamboats, I am greatly surprised at their supineness in allowing one person only [i.e. John Williams] to supply them.[2]

In retrospect the apathy of the land owners towards developing coal mines on their allotments is to some extent understandable. Not only was the market quite small but the portions were so tiny that the coal reserves which each contained did not justify the capital expenditure required for even the most modest operation. Under these circumstances a certain amount of superficial scratching at the outcrop to satisfy personal and intermittent local demand was all that could have been expected and indeed was all that was achieved. The only significant operation to be mounted on either the Woodend or the Tivoli allotments in the twenty year period following their release was Walter Gray's mine, because he had had the foresight to enlarge his holding by purchasing three adjacent portions. All other developments in coal mining occurred elsewhere.

Types of Mines

Outcrops are frequently the result of the natural erosion by wind and rain of the cap or ridge of an anticline. This leads to the exposure of underlying strata whose number and extent tend to increase as the stripping action proceeds. Strata which have been bared by such geological action inevitably dip away from the lines of exposure, and this is true of all but a handful of the measures on the Ipswich coalfield. A second geological action which can produce outcrop formation follows the establishment by a river of a watercourse through a region of relatively horizontal bedding. The strata are then exposed in the banks as erosion deepens the channel, and this is the action which occurred at Redbank and on the Burrum River.

In planning to work a seam exposed at an outcrop the nineteenth century developer had two alternative methods of attack. The first was to tunnel directly into the seam, following the slope of the strata. The second necessitated the location of the seam at some distance back from the outcrop. If it could be found a vertical shaft could be sunk to intersect it and the coal worked towards the outcrop.

Tunnels

A tunnel from the surface had the advantage of requiring no preliminary exploration work and of giving immediate access to coal, although the initial quality of the product was usually greatly inferior to that of the main body of the seam because of the past action of weathering and surface oxidation. Moreover the weathering usually affected adjacent strata tending to make them friable and weak so that the portal to the mine required substantial support in order to keep it open. If the main drive was reasonably level, as at Bundamba, the loaded tubs or wheelbarrows could be man-handled along the tunnel but where the drive followed a gently dipping seam from the portal, as at Redbank, a horse was required to haul the laden tubs up the incline. In cases where the drive was straight and the gradient was steep the horse might be replaced by a cable and steam-driven windlass as at the Old Tivoli Pit. The capacity of the tubs was usually from 250 to 350 kilograms, and in all but the smallest operations they ran on rails (usually constructed of wood) with a gauge of between 0.37 and 0.60 metres.

At Bundamba turntables were fitted at corners to ease the labour of man-handling full tubs out of the mine but at Aberdare, Thomas experimented with an underground bunker situated some sixty metres from the portal which received coal from the workings and loaded it into trucks which were drawn in pairs by horses to the surface. In the early days of the Old Tivoli Pit the tubs were manually pushed onto a large trolley in the main drive which was hauled out by a steam-driven windlass. Later this arrangement was replaced by direct haulage of the tubs in rakes of three which were then coupled directly onto the cable of the tramway on reaching the surface. This procedure had the advantage of utilizing the energy generated by the rake of empty tubs returning underground and reduced reliance on the crude braking mechanism which had previously been employed.

When a tunnel mine stood on or very near the bank of a river, as at Redbank, the underground tramway could easily be extended to a series of bunkers or shoots overhanging the water and usually holding up to one hundred tonnes each. The loaded tubs could then be hauled directly to the shoots ready to deliver their contents into punts, barges or steamers. At intermediate distances from a river a separate tramway was for many years the favoured method of transport. Old Tivoli Pit is such an example. When the mine lay at a considerable distance from water, such as at Aberdare or Bundamba, the owner was forced to rely on horse-drawn drays to transport the coal by road to customers or to the river. This was an expensive and uncertain procedure, severely hampered at many times of the year by the poor state of the roads. The coming of the railway ultimately revolutionized coal haulage from Queensland mines although curiously no serious attempt seems to have been made until the 1880s to lay sidings or connect feeder lines to the main Ipswich to Brisbane line running through the coalfield south of the Bremer River. North of the Bremer River no government railway line was conveniently placed for connecting to any of the mines, and it was necessary to build private tramways, as from Old Tivoli Pit, if road transport was to be avoided.

Shafts

The exploration and development work required for vertical shaft mines was considerably greater than for tunnel mines and a period of up to a year could elapse before any coal was produced. This explains their lack of popularity in Queensland. On the other hand they conferred some distinct advantages over tunnel mines, the main one being that loaded tubs could generally be pushed downhill to the foot of the shaft ready for hoisting instead of being dragged up a gradient which could be as great as one in five. There was also an opportunity for leaving the coal untouched at the outcrop if the quality was poor or if its presence protected the mine from flooding.

The vertical shafts sunk in early Queensland coal mines were invariably less than forty-five metres in depth, oblong in section and timber lined. This followed Scottish[3] rather than English or Welsh coal mining practice where circular brick-lined or cast-iron "tubbed" shafts were usual. Rectangular shafts were probably preferred because they could be more easily lined with timber than circular shafts, and wood was readily available in Queensland. Typical dimensions were 8 feet by 4 feet (2.4 metres by 1.2 metres) and 10 feet by 5 feet (3 metres by 1.5 metres), the section being divided by brattices into three roughly-equal compartments of which two contained balanced cages while the third held a ladderway or acted as an upcast ventilation shaft, particularly in the early stages of mine development. Similar arrangements can be found in small metalliferous mining operations around the world. The shafts were sunk from the surface by boring blast holes with hand-operated drills, charging them with black powder and detonating them with a crude slow-match fuse,[4] sometimes called a miner's squib. This was a hazardous business which frequently led to accidents.[5]

The usual procedure in opening a shaft mine was to sink the shaft with its long axis lying across the dip of the strata. On intersecting the seam, drives were established up dip and cross headings were driven into the coal parallel to the long axis of the shaft although local conditions, the previous experience of the overseer and the current demand for coal were important factors in determining the ultimate plan of the mine.

It was obviously prudent to sink shafts at a sufficient distance from rivers to minimize the risks of inundation. At the same time an attempt was made to site them near enough to water for a wooden-railed tramway to be laid from the pit head to riverside shoots along which the tubs could be hauled by horses (as at Goodna) or pushed by man (as in the case of Walter Gray's mine).

In the smallest shaft mines the coal was hoisted in a box or bucket on a rope attached to a simple hand-operated windlass or horse-whim. Walter Gray's mine used "a machine constructed by Mr. Petrie" which was probably a horse-whim, the capacity of a hand-operated windlass being insufficient for the reported size of the operation.[6] Redbank was the first Queensland mine to install a steam-driven hoisting engine, although the shaft was succeeded by a tunnel entrance before it had seen much use, and Goodna Mine was probably the first shaft mine in the district to rely exclusively on a steam-winder. It had balanced cages and a crude system of chocks or catches for holding the cages at the surface.

Comparative data on a selection of mines opened in South Queensland between 1848 and 1870 are summarized in table 1.

Methods of Working

In both shaft and tunnel mines the coal was won by room and pillar methods.[7] The procedure was to drive headings into the seam at right angles to the main drive at intervals of about thirty-five metres and connect them with rooms or "cut-throughs" lying parallel to the main drive and spaced at intervals of nine or ten metres (figure 10). At Redbank the rooms were made about equal in width to the pillars, giving 50 per cent extraction, but in some mines—notably Goodna, Aberdare and Moggill—the pillars were slimmed down at a later stage giving up to 80 per cent extraction, although a roof collapse at Moggill was blamed on an unsatisfactory pillar-recovery technique. The rooms were normally worked to the rise in order to keep the working places free from seepage and the later operations at Redbank were severely criticized for not following this pattern. Attempts to control the quality of the coal relied on selecting the particular portion of the seam which was to be mined and screening out the finest sizes at the face before loading into tubs. The fines were used to fuel the furnace of the ventilation system and any excess disposed of in the worked-out areas unless they were suitable for feeding to coke ovens. There was no preparation of the coal on the surface.

The coal was won entirely by hand, as indeed it was throughout the world until the 1860s when the first practical mechanical coal cutter was marketed.[8] The normal procedure was for the miner to cut a groove under the coal close to the floor, to a depth of 0.5 to 1.5 metres. The undercut coal was then broken down with a pick, crowbar or wedge. This was hard, slow, uncomfortable and dangerous work. In 1862 Thomas Jones was killed at Redbank while undercutting 5 feet (1.5 metres) into the seam with the object of obtaining a large quantity of coal at one fall. Unfortunately he left the mass of coal unsup-

Table 1. Details of Underground Operations.

Name of Mine	Date of Opening	Date of Report	Type	Main Drive		Haulage		Underground Layout		Number of Miners	Average Output t/day	Comments
				Cross Section	Length	Method	Motive Power	Distance between Headings	Distance between Rooms			
MOGGILL	1848	1854	Tunnel	6ft x 5½ft	60 yds	Wheelbarrow	Man	–	–	–	6	Pillar recovery
WALTER GRAY	1854	1856	Shaft	–	–	Single rails	Man	–	–	–	17	–
REDBANK	1858	1868	Tunnel	7ft x 6ft	700 yds	Single rails	Horse	55 yds	11 yds	25	50 (i)	No pillar recovery 50% extraction
GOODNA	1865	1867	Shaft	16ft x 4½ft	220 yds (ii)	Double rails	Man	55 yds	10 yds	31	65	Pillar recovery 70% extraction
OLD TIVOLI	1866	1867	Tunnel	6ft x 6ft	200 yds	Single rails	Steam-driven Rope	35 yds	–	33	80	Square pillars; rooms initially 5 yards square
BUNDAMBA	1866	1866	Tunnel	6ft x 4ft	80 yds	Single rails	Man	–	–	4	8	Rooms initially 6 yards square
ABERDARE	1870	1871	Tunnel	18ft x 4ft	100 yds	Single rails	Horse	–	11 yds	9	14	Pillar recovery 80% extraction

(i) Figures refer to earlier years than 1868; production had practically ceased by the time that the report was made.

(ii) Seam being worked to the dip at the time of the report.

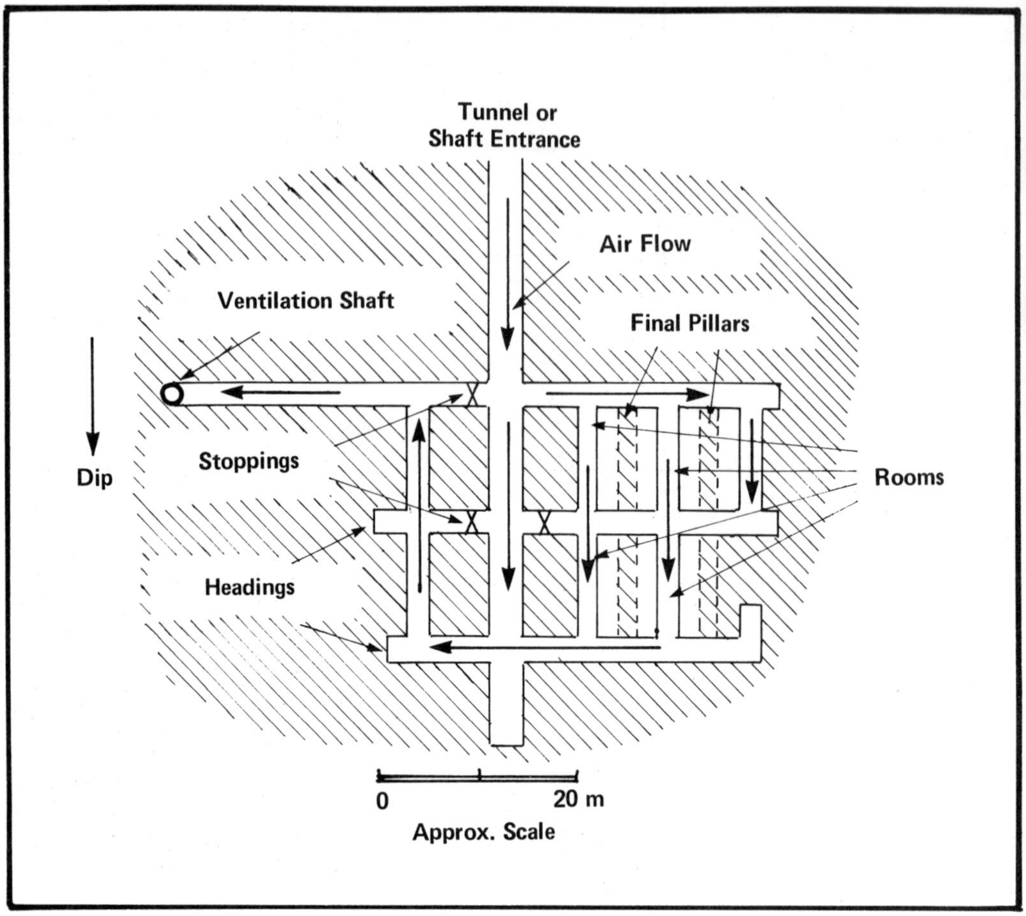

Figure 10. A typical underground working and ventilation plan for a small Queensland coal mine in the 1860s.

ported and a 15 foot (4.5 metre) length fell on him, killing him instantly.[9]

The constant squeezing of the price of coal in the 1860s and the high cost of transporting it by road meant that a mine had to be situated close to a river or a market if it were to operate profitably. From a figure of 20s. per ton ($2 per tonne) in 1860,[10] the price of good-quality coal had fallen by 1871 to 9s. per ton at the pit head,[11] depressing the wages received by the miners. Hewing rates fell to 5s. per ton, with 3s.6d. to 4s. per ton being quite common.[12] Production averaged two tons (tonnes) per man per ten-hour shift in the easily-worked pits, giving a top-class miner a wage of about £9 per month,[13] although the average was generally less than this,[14] particularly as the men were employed on a contract basis[15] and only worked when there were orders to fulfil. The quantity produced by the miner was measured by the number of full tubs he returned to the surface, and the fact that it was not weighed until sold to a customer came in for criticism in the local press.[16]

A major responsibility of the manager or overseer was

to ensure that sufficient faces were open at all times to meet the required production, the practice being to allot a separate face to each miner. On the assumption that on average he could cut, hew and transport two tonnes per shift it was desirable to keep some 250 metres of face open in order to maintain an output of about fifty tonnes per day or three hundred tonnes per week.[17] Work was then available for some twenty-five skilled miners. Production in terms of the output per miner per shift changed very little over the period 1850 to 1870 because it relied entirely on human muscle. Any improvement in productivity which was achieved was met by a demand from the miners' union for the length of the shift to be reduced.

Mines situated close to the Brisbane, Bremer and Burrum rivers were always exposed to the risk of inundation during periods of flood and all suffered this fate at one time or another. Although the mines with portal entrances such as Moggill, Redbank and Walsh's were worst affected neither the shaft mine at Goodna nor Walter Gray's mine at Ipswich escaped unscathed. The direction of dip of the seams did not permit any Queensland mine to use an adit for drainage and some pumping or hoisting arrangements were always necessary to remove water from the workings, although their efficiency varied greatly. Reciprocating force pumps were used at Moggill and Redbank, the former being powered by a horse.[18] Redbank appears to have been the first Queensland mine to install a mechanically-driven pump, the unit being 5 inches (125 millimetres) in diameter with a throw of 5 feet (1.5 metres) and driven by a 10 horsepower (7.5 kilowatt) engine geared down to five strokes per minute.[19] On the other hand Goodna relied on watertight tubs for removing water from the workings and Old Tivoli Pit used a similar technique in its early days.

Firedamp does not seem to have been a problem in most mines until about 1870, both oil lamps and candles being used for illumination.[20] Because the workings were seldom under more than 30 metres of cover, ventilation was easily arranged by sinking small unlined shafts from the surface to appropriate points in the workings (see figure 10) and using them as upcasts by building a furnace at their foot (figure 11). The hoisting shaft or the tunnel entrance then acted as the intake but the arrangements varied very much from mine to mine. In some cases, such as at Moggill, the cover was so thin and the roof so weak that the natural collapse of the ground in the worked-out areas provided sufficient openings to ventilate the workings.[21] In general, a "face airing" procedure was adopted in which the air was conducted successively through each of the working places by constructing "cut-throughs" as and where necessary. With the mines becoming deeper

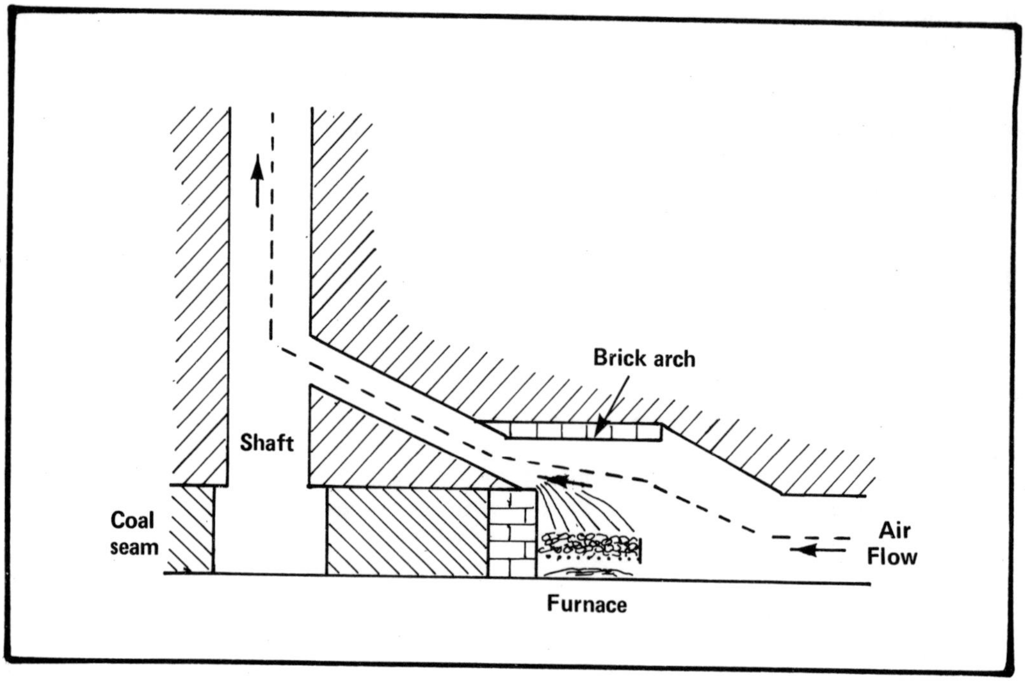

Figure 11. Furnace ventilation in a coal mine.

and more extensive the problems of ventilation increased but there were no rules, regulations or codes of practice to guide the manager. A firedamp explosion at Old Tivoli Pit in 1874 which severely injured a couple of miners, occurred while they were fixing safety lamps in the vicinity of an old shaft so that the dangers of gas must have been realized.[22] Matters came to a head in 1877 when the coal miners petitioned the Legislative Assembly for the enactment of safety regulations and the appointment of an inspector to enforce them (document 11), a special point being made of the inadequacies and dangers of existing ventilation practices.[23]

Safety Legislation

Statutory safety requirements for mines were virtually non-existent until the 1880s and there is no evidence that the 1854 Mines Registration and Inspection Act was ever applied to the Northern Districts either before or after Separation, even though it required nothing more from an owner than the lodging of the plans of any underground workings with the examiner. The first New South Wales examiner, William Keane, was appointed in 1854 but it does not appear that he ever visited his northern outposts before Separation. After Separation the Act was embodied in the Queensland Statutes and nominally remained in force until the comprehensive Mines Regulation Act reached the Statute Book in 1881, but no examiner ever

1877.

QUEENSLAND.

Legislative Assembly.

QUEENSLAND COAL-MINES.

(PETITION.)

Printed under direction of the Printing Committee.—See Eleventh Report.

To the Legislative Assembly of Queensland.

We, the Coal-miners assembled for the purpose hereinafter expressed, beg to petition the Honourable Assembly.

TIME ARRIVED FOR LEGISLATION.

Your petitioners show that the time hath arrived when the Queensland coal-mines should be regulated by a special code of rules and regulations adapted to coal-mining, and that a practical coal-miner be appointed inspector of coal-mines.

IMPORTANCE OF LEGISLATION.

Your petitioners show that such rules and regulations are of very grave importance and necessity, enforced by a practical coal-miner as inspector. The gravity of the importance becomes evident. In the first place—Ventilation to the coal-miner is of *vital* importance. If it be defective, the miner's energy and strength become prematurely enfeebled, and enervation and pallidness ensue, and an untimely death follows, lamented by a widow and orphans. He is liable to many other perils to life, and grievances deteriorating to health, which, by proper precaution by those responsible, enforced by legislative regulations, may be averted.

PROPRIETORS OF COAL-MINES SUFFER.

In the second place—The proprietors of coal-mines suffer great losses when these defects prevail, hundreds of pounds are unnecessarily spent. The coal-miner cannot render a full day's work where defective ventilation, perils to life, and deteriorating grievences prevail. Moreover, practical coal-miners from the old country do not remain long in such mines; they clear out, and take to any other labour, or emigrate to New South Wales, where they can be employed under Government inspection and legislative regulations. Therefore, the employers are compelled to employ unskilled labour, to their own pecuniary loss, and to the loss of trade, in consequence of the inferior production of such labour.

IMPORTANCE TO THE GOVERNMENT OF QUEENSLAND.

In the third place—Your petitioners show that it is of *paramount* importance to the Government of the Colony of Queensland, because each skilled coal-miner costs the Government a large sum of money, probably not less than £100, for his individual importation into the colony; and if he have a family, the cost may be doubled or trebled. When the skilled miner finds that the coal-mines in Queensland are not under Government inspection, and that there are no legislative coal-mine regulations, he leaves the colony as soon as he can save enough money to pay his passage to Newcastle. The skilled miners who left Queensland within this year, 1877, are a dead loss of thousands of pounds to the Government.

For these reasons, therefore, your petitioners would urge the Honourable Assembly to appoint, without delay, a practical coal-miner to be inspector of the coal-mines of Queensland, and that a legislative code of rules and regulations be issued forthwith. And your petitioners will ever pray.

Signed on behalf of a public meeting of coal-miners,

EDWARD GITTINS,

Chairman.

Tuesday, October 2nd, 1877.

(*Here follow* 150 *Signatures.*)

[*Price 3d.*]

By Authority: JAMES C. BEAL, Government Printer, William street, Brisbane.

Document 11. Petition presented by the coal miners to the Legislative Assembly in 1877 (chap. 12, n. 23).

appears to have been appointed. It was established during the course of the *Towns* v. *Campbell* legal case in 1867 that no plans had ever been made of the Redbank Mine, the overseer, John Corbett[24] averring that they were quite unnecessary because he knew the workings intimately and there was no other colliery adjacent to Redbank with whom their workings could interfere! The fallacy of the argument was exposed some years later when it was decided to start a new mine at Redbank and there was complete uncertainty regarding the exact location of the old worked-out area.

In Britain in 1862 it became illegal to operate a mine which relied on a single shaft for access to the workings; this followed the Hartley disaster in Northumberland where the one shaft became blocked and 204 miners lost their lives.[25] Single-entrance mines continued to be accepted in Queensland, and the strengthening in New South Wales of the 1854 Mines Registration and Inspection Act by important amendments in 1862 again made no impression on Queensland's legislators. The need for a royal commission to inquire into all aspects of mining in the colony was pointed out in 1872[26] and the appointment of inspectors of mines was suggested two years later.[27] When Thomas McIlwain was named as the first minister for mines in 1874, George Lukin was made the colony's first inspector of mines,[28] but his main task was to establish and administer the warden system in the goldfields rather than establish safer mining procedures. It was not until the appointment of William Fryar as inspector of mines and the passage of the Mines Regulations Act in 1881 that the enforcement of the most elementary safety measures became possible.

Geology

The systematic development of coal mining was seriously hampered in many cases by an absence of knowledge of the detailed geology of the coalfields. Provided that adequate supplies of suitable coal could be won from seams outcropping adjacent to a navigable river there was no real problem. The difficulties began when such reserves were lost through inundation or when a high quality seam was exhausted or could not be traced beyond a geological fault. The outcropping seams on the original Tivoli allotments and at Redbank returned ash assays of at least 12 per cent which was double the value of the best Newcastle coals. "Fryar's Seam" at Moggill was particularly valuable because it yielded the lowest ash of all the local coals until the discovery of the Aberdare and Tivoli seams in the mid 1860s, and in 1858 it commanded a 50 per cent premium over Ipswich and Redbank products because of its quality.[29] The major reason for the transfer in 1858 from Brisbane to Newcastle of the coaling stop for the steamboats of the Australasian Steam Navigation Company operating the Sydney to Brisbane service was the uncertainty of supply of Moggill coal, although a contributing factor was undoubtedly the financial interest which the company had acquired in the Newcastle mines.[30] The numerous attempts to establish a new entrance to Moggill Mine which would be less susceptible to river flooding than the original and might also give access to the seam beyond the fault were all unsuccessful. They failed because the geology of the area was neither recognized nor understood and the true situation was not appreciated for many years. It was unfortunate that in this case the geological structures were particularly complex. Reference to the map of the area prepared by the Queensland geological survey[31] in 1955 shows a markedly different picture of the location of the Moggill seams from the 1899 or the 1907 geological maps[32] (figure 12). Bearing in mind that the two latter maps, with all their inaccuracies, embody data collected over a period of almost fifty years following the closure of Moggill Mine, the difficulties facing the coal prospectors in 1860 who had only their eyes to rely on can well be appreciated.

Prospecting for coal away from an outcrop usually relied on the excavation of small shafts. The alternative of drilling exploratory holes met with only limited acceptance, probably because it depended entirely on manual labour until about 1870 when steam-driven drills were introduced.[33] A typical procedure adopted was described by an observer in 1872.

> Operations are commenced with a common two-inch auger, fitted on to an iron shaft or rod, about six feet long, and about three-quarters of an inch square. This is first stuck in the ground, and two men begin to bore down, exactly in the same way as if they were boring a log of timber, drawing their instrument out as often as the screw is filled with the clay or other strata they may pass through. Very hard stuff can be penetrated by this simple process, and for a depth of ten or twelve feet nothing more is required. When, however, considerable depths have to be reached, a framework of saplings is erected over the box—a long pole with a pulley and windlass to pull up the rods from fifteen to thirty feet long, which are screwed on in successive lengths as the bore descends. When the auger encounters rock too hard for the screw to penetrate, a drill-head is screwed on in its place and the bore descends by jumping . . . The handle at the top of the rods is hung from the end of a long sapling, which gives the necessary spring to enable the workmen to lift them, and they fall again with an additional force given them by the borers, who keep lifting them and dropping them as they slowly turn them round in a continuous circle. On pulling up the rods, the rock being penetrated is found in small grains, coating the point of the drill, and can be easily separated by washing from the

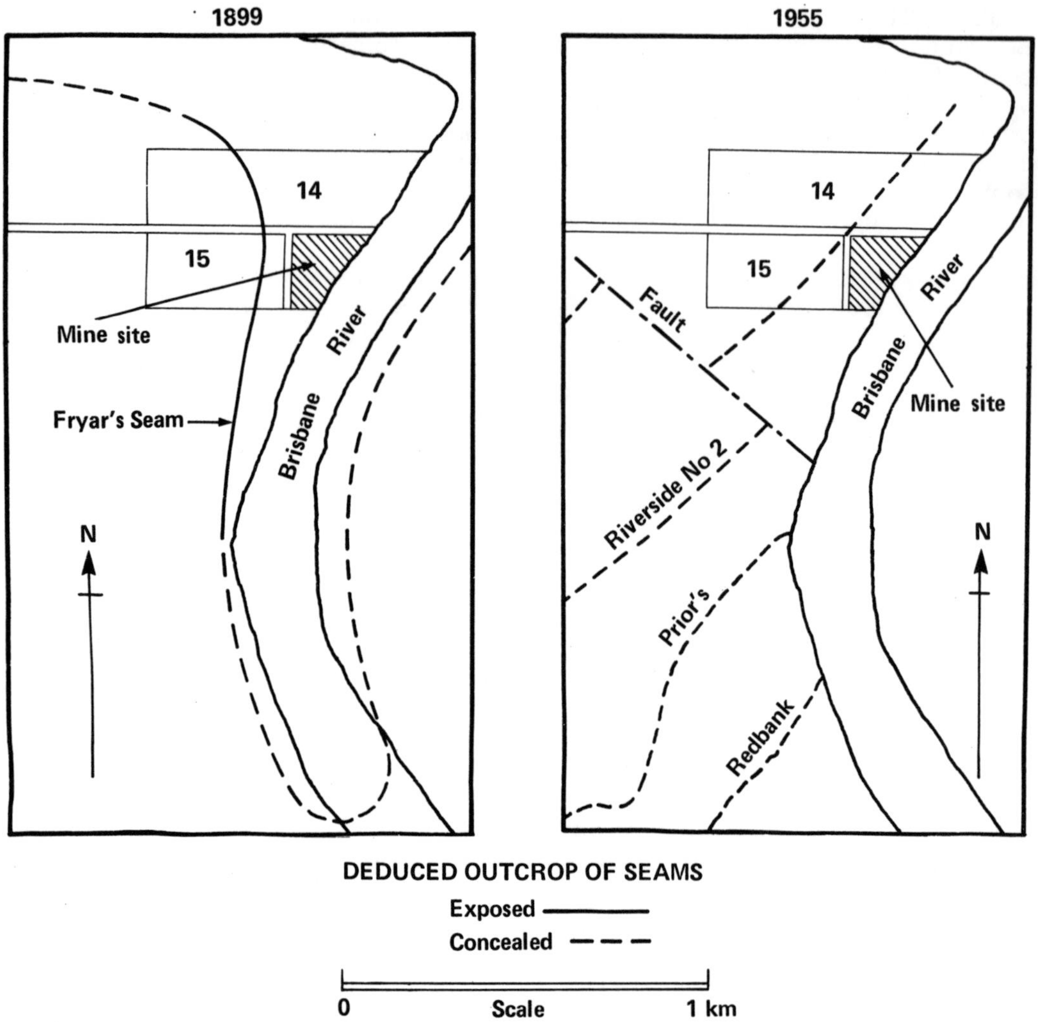

Figure 12. Geology of the Moggill area as deduced by Cameron in 1899 (chap. 12, n.32) and by Denmead in 1955 (chap. 12, n. 31).

> mud and clay with which it is covered . . . The rate of progress is very rapid in favorable soils, such as clays and soft rocks, as much as thirty-two feet having been penetrated in one day . . . penetrating very hard rock at the . . . rate of three feet a day . . . The rods and ironwork of the apparatus we saw, capable of penetrating to a depth of 200 ft, would cost about £25; and any bushman, with a few yards of rope, greenhide, and trace chain, could put up the rest.[34]

Government involvement in the geology of Queensland dates from 1867 when the colony was divided into two districts for the purpose of survey. Richard Daintree was appointed to the Northern District, and D'Oyley Aplin to the Southern, the Queensland geological survey being established two years later.[35] Daintree defined the objective of the survey with great clarity: "In a newly settled Colony the first objective of the geologist is the determination of those districts in which the best land or available mineral wealth is to be obtained, accuracy in detail and purely scientific investigation being deferred to an often remote future''.[36] One of the earliest maps showing the location of coal was prepared by Daintree in 1873 after he had been made agent general in London, and it indicates a remarkable understanding of the general distribution of coal in Queensland (figure 13).[37] By that time

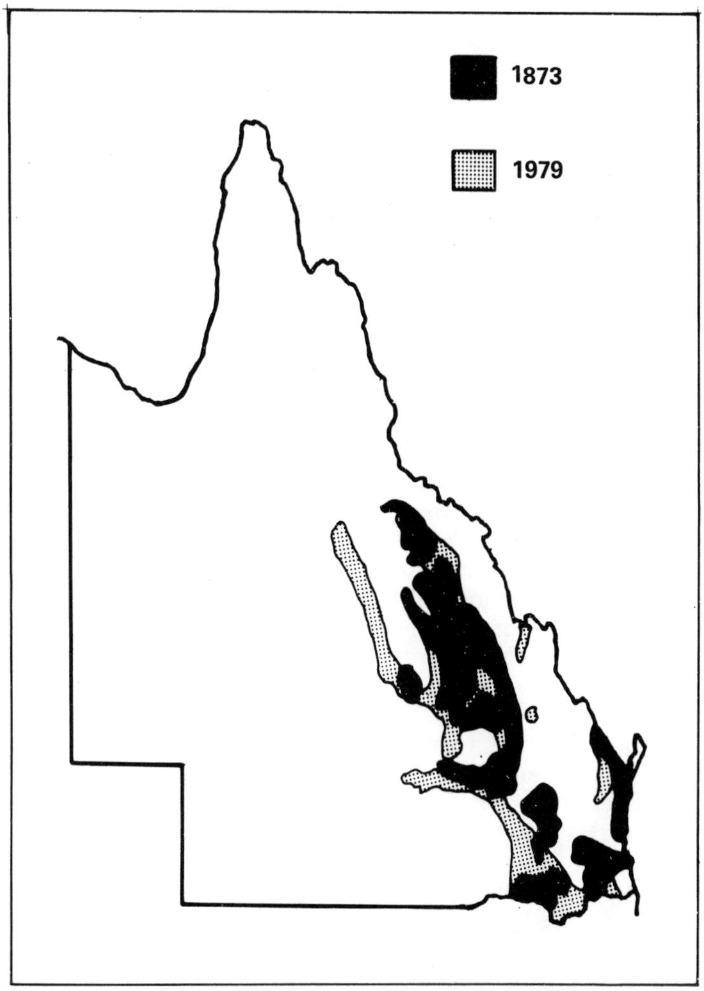

Figure 13. Queensland's coalfields as recorded by Daintree in 1873 (chap. 12, n. 37), compared with a 1979 appreciation by the Queensland Government Mines Department.

coal had been located, or was being mined in a small way at Bingera (behind Bundaberg), on the Pelican Creek near Collinsville, down the Mackenzie and Isaac rivers, and as far west as Clermont in addition to the Brisbane Valley and the Ipswich area, the Darling Downs, Rosewood and the Burrum River. It was the detailed picture which was much less obvious and Daintree suggested that a company which had access to the newly-patented technique of diamond drilling should be commissioned to complete 1000 feet (300 metres) bores at the following four points in the supposed coal measures:

1. In the Moreton district, on the line of the railway in the neighbourhood of Ipswich.
2. In the Darling Downs district, on the line of the railway between Toowoomba and Warwick, say Allora.
3. On the Mary River where coal measures occur most favourable for shipment.
4. On the Fitzroy coal-field most available for making use of the extension of the Northern railway from Rockhampton.[38]

There was no government support for this proposal and, with the exception of a few geological notes on the West Moreton coal measures which were published in 1883 by Tenison-Woods,[39] there was no reliable information available on the structure of any of the Queensland coalfields until the turn of the century.[40]

Coke Making

The trivial requirements of the early Moreton Bay settlement for charcoal for foundry work was met by occasional charcoal burning in the local luxuriant forests. However the conveniently-situated timber was soon consumed for general fuel or constructional work and the rising demand from various sections of the community for charcoal—or its equivalent—encouraged the search for alternatives. As early as 1843 it was reported that Ipswich blacksmiths were using local coal in their forges[41] and any discussion of the properties of newly-discovered seams almost invariably included a comment on their caking qualities.

One of Queensland's best coking coals came from the Tivoli seam in North Ipswich and it was first mined in a substantial way in 1867 by Hooper and Robinson. The decision only two years later by the Southern and Western Railway to fuel its locomotives with coke instead of billet wood provided the impetus for the partners to construct Queensland's first coke oven battery, although they may already have been contemplating the erection of a few ovens to supply local demands.

Before the invention of the by-product process in the 1860s in Europe (and for many years afterwards in parts of the world where production was too small to justify the collection of the by-products), the type of oven universally employed was that known as the beehive.[42] It consisted of a fireclay or brick chamber, beehive-shape in cross-section and circular in plan, provided with a charging door at ground level and with an exit flue in its crown for carrying away the waste gases. The ovens were arranged in parallel rows, set in an outer rectangular structure of stonework or masonry with the intervening space filled with earth or fireclay in order to provide heat insulation for the individual units (figure 14).

Each row, or battery, of ovens was charged and emptied according to a set pattern. Coal was shovelled into an empty oven to a height of 0.6 metres or so and ignited. The door was bricked up or closed with an iron plate and plastered except for a small hole some 150 millimetres square which was left at the top of the temporary closure for the admission of air. The upper layer of the coal burnt and initiated the distillation of volatile material from the lower layers, the evolved gases burning in the dome of the ovens as they met the air supply drawn in through the top of the charging door. The dome was soon heated to a high temperature and greatly assisted in carbonizing the charge by radiating its heat to the coal. The operation of distillation and immediate combustion proceeded until the whole of the volatile matter in the coal had been evolved, which occupied about three days. The coke was then quenched by inserting a water sprinkler through the air hole in the door and was withdrawn manually with rakes. The volatile products were not always completely burnt in the dome and at certain parts of the coking cycle it was usual for the ovens to emit clouds of dense and pungent smoke. Moreover it was impossible to avoid burning a certain proportion of the coke in the process so that almost two tonnes of coal were required to produce one tonne of coke.

The ovens which were constructed at Old Tivoli Pit in 1869 followed the almost universal practice of being 10 feet (3 metres) in internal diameter and taking a charge of approximately five tonnes each (figure 14). One oven could accept two charges per week giving it an output of some six tonnes of coke and providing a maximum production by the battery of twenty-three ovens of 120 to 140 tonnes per week.

Coke making procedures hardly changed at all in the Ipswich area during the period 1870 to 1958. Production finally ceased because of the labour-intensive nature of the operation and the atmospheric pollution inherent in the process.

The Coal Miners

Of the men who owned or leased the mines we know a little; of those who worked in them we know much less. The colonial census for 1861 gives the number of coal miners employed in Queensland as a mere 29 but this figure rose quickly, reaching 74 in 1864 and 122 by 1871[43] without a comparable increase in production, so far as

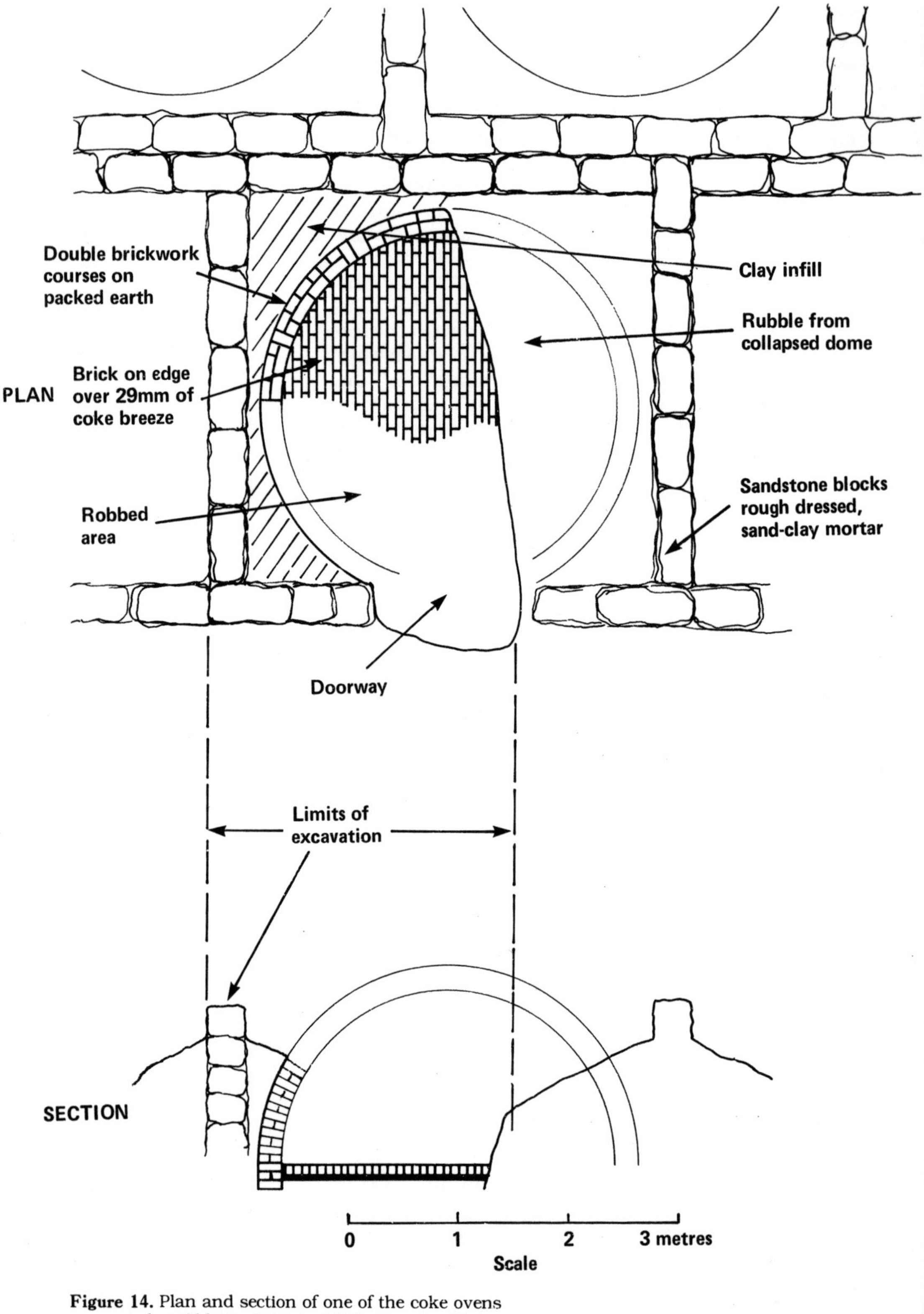

Figure 14. Plan and section of one of the coke ovens excavated at Old Tivoli Pit in 1979. See also Plates 28–30, pp. 122–24.

can be judged from the incomplete statistics which are available. It is impossible to discover how many were in employment at any time but it is known that they worked on a piece-work or contract basis.[44] Their working conditions were described by a visitor:

> Men naked to the waist, and almost as black as niggers, working away in every conceivable attitude in remote corners, others filling waggons and running them out to the stage in readiness for the trolly, others screening "slack", and some taking a spell. What was even more interesting than this was to see them partaking their mid-day meal, for then they all assembled on the stage, and presented a laughable picture to the visitor. A glance around, however, is sufficient to banish all mirthfulness, for the dread of some accident which might result in serious consequences is rarely absent from a person unaccustomed to such places, and every unusual sound causes alarm. A very pleasant feeling of relief is experienced on reaching the surface of the earth.[45]

Despite the growing counter-attraction of gold mining in Queensland from 1860 onwards there is no evidence to suggest that there was any serious shortage of coal miners and the opposite is probably more nearly the truth.[46] Gold mining apparently had a different kind of appeal to coal mining.

The early mine proprietors—such as John Williams, Walter Gray, John Ferrett, John Campbell, Harry Hooper and William Walsh—were generally pastoralists, settlers or commercial adventurers with interests in a range of enterprises to which coal mining made only a minor contribution. But although their mines were small and operated at shallow depth they still required some experienced supervision to maintain production, particularly after the exposed outcrops had been exhausted. Most of these skilled workers came from the United Kingdom by way of Newcastle in New South Wales. William Simpson who was the overseer at Redbank in 1861 had worked in Scotland for eleven years as a miner before emigrating to Australia,[47] and his successor John Corbett had been actively engaged in coal mining in Lanarkshire and Newcastle (New South Wales) for the previous quarter of a century.[48] Robert Towns's nominee for overseer of the Redbank Mine in 1867 was George Curless who claimed thirty-five years practical experience, beginning in the north of England and including ten years as a deputy coal overseer in Newcastle, New South Wales.[49] He had been a leading trades unionist during the strikes and agitation in 1861, when he was acclaimed a "popular and very useful man" whose pleas for moderation were a major factor in defusing what could have been an ugly situation.[50]

The only mining engineers of the period who have so far been identified with any certainty are William Fryar who was manager of Moggill Mine in the late 1850s, Thomas

Pearson Moody who for a short period from 1869 took over responsibility for the management of the Redbank estate[51] which included Goodna Mine, and James Robertson whose main activities were confined to the Burrum River coalfield. Fryar was the son of an English mining engineer and possessed "scholastic attainments"[52] of which we unfortunately know nothing, but it is fair to assume that most of his expertise was acquired while working on the job. Moody was spoken of with great respect during the *Towns* v. *Campbell* case and later became manager of the Waratah Colliery in New South Wales.[53] Robertson, who was called as an expert witness by the defence in the Redbank case claimed to be a "duly qualified Mining Engineer"[54] and later managed his own mine at Torbanlea on the Burrum River.[55] He does not appear to have held any formal mining qualifications and relied on four years' experience which he gained with his brother as a surveyor for the Clyde Iron Company, and five years as the proprietor of a small pit in Scotland before emigrating to Queensland in 1865.

In the ancillary operations involved in opening up some of the mines the activities of the early engineers, such as John Petrie and William Pettigrew, are mentioned but there is insufficient detail to evaluate the extent of their contributions or the kind of work which they may have done. What is clear is that two of the chief operating mines of the period—Moggill and Redbank—were incredibly badly run, even by the standards of the time, reflecting the abysmal ignorance and poor quality of the supervising staff. Durisdeer Mine on the Mary River failed to produce a single tonne of coal in its short existence because of errors in the development stage. Most miners with a taste for risk or adventure forsook coal mining for the heady excitement of the gold, copper and tin rushes which were a feature of the 1860s and 1870s in Queensland.

But changes were in the wind. In the second half of the 1860s Lewis Thomas launched into colliery proprietorship, bringing to it a blend of practical experience, hard work and business acumen which was to have a profound influence on the Queensland coal trade over the next fifty years and lift him to a position of power and prestige in the community. In North Ipswich James Gulland joined the partnership of Hooper and Robinson in 1870 and his efforts were soon felt throughout the infant industry. The Staffords, the Wrights and other adventurers were waiting in the wings. These entrepreneurs located and tapped coal measures whose quality was greatly superior to that of Redbank or Goodna and rivalled Newcastle's best products. Nevertheless full exploitation still had to await the provision of suitable transport and loading facilities which were essential to an export trade. The impetus came with the opening of the railway line between Ipswich and Brisbane in 1874, exactly fifty years after the first discovery of coal in Queensland, leading some years later to the establishment of coal handling facilities on the river bank at South Brisbane and the development of overseas markets.

Coal mining was a minor, subordinate activity in Queensland in the middle of the last century. Although the potential of the known coalfields had been recognized for many years the time was not ripe for their development. The situation had been summed up by the government statistician in his first report to the Queensland parliament in 1860. The introduction includes the following comment: "Little attention has yet been paid [to coal mining] beyond sinking a few shafts into one or two of the most accessible coal seams for the supply principally of steam boats plying on the River Brisbane".[56]

But at least it was a start, and from insignificant beginnings grew the West Moreton coalfield, to be followed a century later by the vast expansion in central Queensland which to-day is turning the state into a major coal producer on a world scale.

Notes and References for Chapter 12.

1. *Statistics of New South Wales*, 1870; and 1871.
2. *Votes and Proceedings of the Legislative Council of New South Wales*, 1854, Vol.2, pp.1383–95.
3. M. Dunn, *Treatise on the Winning and Working of Collieries*, 2nd ed. (London: Simkins and Marshall, 1852).
4. J.A.S. Ritson, *A History of Technology*, ed. C. Singer, E.J. Holmyard, A.R. Hall and T.I. Williams, 4th ed. (Oxford: Clarendon Press, 1958), p.64.
5. *Moreton Bay Courier*, 13 February 1858.
6. R.S. Lewis, *Elements of Mining*, 2nd ed. (New York: J. Wiley & Sons Inc., 1941).
7. The method is also referred to as pillar and stall, bord and pillar, and stoop and room depending on the locality of the mine.
8. H.F. Bulman and R.A.S. Redmayne, *Colliery Working and Management*, 2nd ed. (London: Crosby, Lockwood & Son, 1906).
9. *Brisbane Courier*, 20 August 1862.
10. *Moreton Bay Courier*, 6 October 1860.
11. *Votes and Proceedings of the Legislative Assembly of Queensland*, 1871, pp.675–837.
12. *Moreton Bay Courier*, 14 June 1861.
13. *Votes and Proceedings of the Legislative Assembly of Queensland*, 1878, Vol.2, pp.319–402.
14. *Queensland Times*, 19 October 1867.
15. *Votes and Proceedings of the Legislative Assembly of Queensland*, 1871, pp.675–837.
16. *Queenslander*, 28 October 1871.
17. *Towns v. J. and J.E. Campbell*, 1867. Queensland State Archives. Accession 25/91. SCT.37 (George Curless's Testimony).
18. *Moreton Bay Courier*, 5 February 1859.
19. *Towns v. J. and J.E. Campbell*, 1867. Accession 25/92. SCT.38 (James Robertson's Testimony).
20. *Queensland Times*, 28 October 1871; and 19 October 1867.
21. *Votes and Proceedings of the Legislative Council of New South Wales*, 1854, Vol.2, pp.1383–95.
22. *Queenslander*, 2 May 1874.

23. *Votes and Proceedings of the Legislative Assembly of Queensland*, 1877, Vol.3, p.245.
24. *Towns* v. *J. and J.E. Campbell*, 1867. Accession 25/92. SCT.38 (John Corbett's Testimony).
25. R. Nelson Boyd, *Coal Mines Inspection* (London: W.H. Allen & Co., 1879).
26. *Queenslander*, 7 September 1872.
27. Ibid., 7 November 1874.
28. *Queensland Blue Book*, 1874.
29. *Statistics of New South Wales*, 1858.
30. *Moreton Bay Courier*, 1 March 1860.
31. A.K. Denmead, "The West Moreton (Ipswich) Coalfield", *Queensland Geological Survey*, Publication 279 (Brisbane: Government Printer, 1955).
32. W.E. Cameron, "Geology of the West Moreton or Ipswich Coalfield", *Queensland Geological Survey*, Publication 147 (Brisbane: Government Printer, 1899); and Publication 204, 1907.
33. *Queensland Times*, 21 January 1871.
34. *Queenslander*, 5 October 1872.
35. *Queensland Blue Book*, 1869.
36. R. Daintree, "Notes on the Geology of the Colony of Queensland", *Quarterly Journal Geological Society of London* 28 (1872): 271–360.
37. R. Daintree, *Queensland, Australia* (London: G. Street, 1873).
38. *Queenslander*, 1 November 1873.
39. J.E. Tenison-Woods, *The Coal Resources of Queensland* (Brisbane: James C. Beal, Government Printer, 1883).
40. Cameron, "Geology of the West Moreton or Ipswich Coalfield".
41. *Sydney Morning Herald*, 22 August 1843.
42. J.F. Thorpe and M.A. Whiteley, *Thorpes Dictionary of Applied Chemistry*, 4th ed, Vol.3 (London: Longmans, Green & Co., 1939), pp.258–61.
43. *Statistics of Queensland*, 1861; 1864; and 1871.
44. *Queenslander*, 18 June 1870.
45. Ibid.
46. *Votes and Proceedings of the Legislative Assembly of Queensland*, 1878, Vol.2, pp.319–402.
47. *Towns* v. *J. and J.E. Campbell*, 1867. Accession 25/90. SCT.36 (William Simpson's Testimony).
48. *Queenslander*, 2 May 1874.
49. *Towns* v. *J. and J.E. Campbell*, 1867. Accession 25/91. SCT.37 (Curless).
50. *Sydney Morning Herald*, 2 October 1861.
51. *Votes and Proceedings of the Legislative Assembly of Queensland*, 1871, pp.675–837.
52. *Queensland Government Mining Journal* 5 (1904): 309.
53. *Sydney Morning Herald*, 17 January 1873.
54. *Towns* v. *J. and J.E. Campbell*, 1867. Accession 25/92. SCT.38 (Robertson).
55. *Votes and Proceedings of the Legislative Assembly of Queensland*, 1900, Vol.2, pp.441–42.
56. *Statistics of Queensland*, 1860.

Select Bibliography

Official Sources

Votes and Proceedings of the Legislative Council of New South Wales
Votes and Proceedings of the Legislative Assembly of Queensland
Queensland Parliamentary Debates
Queensland Government Gazettes
Queensland Blue Books
Queensland Geological Survey, *Publications and Reports*
Mines Department, Government of Queensland; *Annual Reports*
Titles Office, Government of Queensland
Department of Mapping and Surveying, Government of Queensland
Queensland Government Mining Journal
Statistics of New South Wales
Statistics of Queensland

Principal Government Reports consulted

New South Wales

1847, *Select Committee on Coal Enquiry*
1850, *Select Committee on Steam Communication with Europe*
1852, *Select Committee on Steam Navigation Bill*
1854, *Report on Condition of "Shamrock" on her last Return from Moreton Bay*
1854, *Twelfth and Thirteenth Trimonthly Reports* by S. Stutchbury
1855, *Report on Steam Communication with England*

1855, *Report on Navigation on the Rivers Brisbane and Bremer*

Queensland

1865, *Select Committee on Existing and Proposed Lines of Steam Communication*
1865, *Conditional Grants of Land for Mining Coal*
1867, *Evidence Relative to the First, Second, Third and Fifth Sections of the Southern and Western Railway*
1871, *Supply of Coke for Railway Purposes*
1871, *Commission into Proposed Extension of the Southern and Western Railway from Ipswich to Brisbane*
1876, *Coal Deposits of the West Moreton and Darling Downs District*
1876, *General Working of Railway Department for Year Ending 31 December 1875*
1876, *Supply of Coal for Southern and Western Railway*
1878, *Royal Commission into the Best Route for Extending the Southern and Western Railway to Navigable Water*
1879, *The Burrum Coal Mines*
1880, *Prospectus of the Queensland Burrum Coal and Railway Company (Limited)*
1881, *Select Committee on Gulland Tramway*
1900, *Royal Commission on Accidents in Coal Mines*

Manuscripts and Photographs

Mitchell Library, Sydney. (Especially King Papers and Microfilms A227/632, 637).
New South Wales Archives. (Especially Location 4/2460.3, Ref. 37/10899).
John Oxley Library, Brisbane. (Especially photographs).
Queensland State Archives. (Especially Accessions 25/90-92, Ref. SCT 36-38).

Journals and Newspapers

Royal Australian Historical Society (and *Australian Historical Society)*
Royal Queensland Historical Society (and *Historical Society of Queensland)*
Maryborough Chronicle
Moreton Bay Courier (and *Brisbane Courier*)
North Australian
Queensland Daily Gazette
Queensland Times

Queenslander
Sydney Gazette
Sydney Morning Herald

Printed Books and Papers

Atkinson, F. *The Great Northern Coalfield.* London: University Tutorial Press, 1968.

Australian Dictionary of Biography, vols. 1-6. Melbourne: Melbourne University Press, 1967-1976.

Barnard, A. *The Australian Wool Market, 1840-1900.* Melbourne: Melbourne University Press on behalf of Australian National University, 1958.

Bateson, C. *Patrick Logan.* Sydney: Ure Smith, 1966.

Bolton, G.C. *A Thousand Miles Away.* Canberra: Australian National University Press, 1970.

Boyd, R.N. *Coal Mines Inspection.* London: W.H. Allen, 1879.

British Parliamentary Papers (Australia). Shannon: Irish University Press, 1970.

Bulman, H.F. and Redmayne, R.A.S. *Colliery Working and Management.* 2nd ed. London: Crosby, Lockwood & Son, 1906.

Coghlan, T.A. *Progress of Australia in the 19th Century.* Sydney: Government Printers, 1903.

Clark, Manning. *A Short History of Australia.* London: Heinemann, 1969.

Daintree, R. *Queensland, Australia.* London: G. Street, 1873.

Dunn, M. *Treatise on the Winning and Working of Collieries.* 2nd ed. London: Simkins & Marshall, 1852.

Ellis, M.H. *A Saga of Coal.* Sydney: Angus & Robertson, 1969.

Fitzpatrick, B. *The British Empire in Australia.* Melbourne: Melbourne University Press, 1941.

Fox, M. *The History of Queensland.* Brisbane: State Publishing Co., 1919.

Gregson, J. *The Australian Agricultural Company, 1824-1875.* Sydney: Angus & Robertson, 1907.

Hamilton, H. *History of the Homeland.* London: George Allen & Unwin Ltd, 1947.

Historical Records of Australia, series I, vols. 1-46. Canberra: Library Committee of Commonwealth Parliament.

Jose, A.W. and Carter, H.J. *The Australian Encyclopaedia.* Sydney: Angus & Robertson, 1925.

Knight, J.J. *In the Early Days.* Brisbane: Sapsford & Co., 1895.

Leichhardt, L. *Journal of an Overland Expedition in Australia.* London: P. & W. Boone, 1847.

Lewis, R.S. *Elements of Mining.* 2nd ed. New York: J. Wiley & Sons Inc., 1941.

McClurg, J.H.C. *Historical Sketches of Brisbane*. Brisbane: Library Board of Queensland and Royal Historical Society of Queensland, 1975.

McKellar, N.L. *From Derby Round to Burketown*. St Lucia: University of Queensland Press, 1977.

Meston, A. *Geographic History of Queensland*. Brisbane: Government Printer, 1895.

Petrie, C.C. *Tom Petrie's Reminiscences of Early Queensland*. Brisbane: Watson Ferguson, 1904.

Pugh's Almanac. Brisbane: T.P. Pugh and Thorne & Greenwell.

Raistrick, A. *Industrial Archaeology*. London: Paladin, 1973.

Ross, J. *A Treatise on Navigation with Steam*. London: Longman, Rees, Orme, Brown & Green, 1828.

Russell, H.S. *The Genesis of Queensland*. Sydney: Turner & Henderson, 1888.

Singer, C., Holmyard, E.J., Hall, A.R. and Williams, T.I., eds. *A History of Technology*. 4th ed. Oxford: Clarendon Press, 1958.

Steele, J.G. *The Explorers of the Moreton Bay District, 1770–1830*. St Lucia: University of Queensland Press, 1972.

Steele, J.G. *Brisbane Town in Convict Days, 1824–1842*. St Lucia: University of Queensland Press, 1975.

Tenison-Woods, J.E. *The Coal Resources of Queensland*. Brisbane: James C. Beal, Government Printer, 1883.

Thomson, A.K. *The Collected Works of Thomas Wesley*. Brisbane: Jacaranda Press, 1967.

Thorpe, J.F. and Whiteley, M.A. *Thorpes Dictionary of Applied Chemistry*. 4th ed. London: Longmans, Green & Co., 1939.

Waterson, D.B. *A Biographical Register of the Queensland Parliament, 1860–1929*. Canberra: Australian National University Press, 1972.

Index